Lecture Notes in Computer Science 12340

More information about this series at http://www.springer.com/series/7407

Minming Li (Ed.)

Frontiers
in Algorithmics

14th International Workshop, FAW 2020
Haikou, China, October 19–21, 2020
Proceedings

 Springer

Editor
Minming Li
City University of Hong Kong
Hong Kong, China

ISSN 0302-9743   ISSN 1611-3349 (electronic)
Lecture Notes in Computer Science
ISBN 978-3-030-59900-3   ISBN 978-3-030-59901-0 (eBook)
https://doi.org/10.1007/978-3-030-59901-0

LNCS Sublibrary: SL1 – Theoretical Computer Science and General Issues

This Springer imprint is published by the registered company Springer Nature Switzerland AG
The registered company address is: Gewerbestrasse 11, 6330 Cham, Switzerland

Preface

This volume contains the papers presented at the 14th International Frontiers of Algorithmics Workshop (FAW 2020), held during October 19–21, 2020, at Hainan University in Haikou, China. Due to COVID-19, the whole conference was conducted online. In order to conduct a real-time, interactive online conference, we arranged three half-day programs where each day started with a keynote talk, followed by one session consisting of four talks.

The workshop provides a forum on current trends of research in algorithms, discrete structures, and their applications, and brings together international experts at the research frontiers in these areas to exchange ideas and to present new results.

The Program Committee, consisting of 30 top researchers from the field, reviewed 15 submissions and decided to accept 12 papers. Each paper had three reviews, with additional reviews solicited as needed. The review process was conducted entirely electronically via EasyChair. We are grateful to EasyChair for allowing us to handle the submissions and the review process, and to the Program Committee for their insightful reviews and discussions, which made our job easier.

The best paper goes to "Minimizing Energy on Homogeneous Processors with Shared Memory," authored by Vincent Chau, Chi Kit Ken Fong, Shengxin Liu, Elaine Yinling Wang, and Yong Zhang.

Besides the regular talks, the program also included three keynote talks by Evripidis Bampis (Sorbonne Université, France), Xujin Chen (Chinese Academy of Sciences, China), and Kazuhisa Makino (Kyoto University, Japan). We are very grateful to all the people who made this meeting possible: the authors for submitting their papers, the Program Committee members and external reviewers for their excellent work, and the three keynote speakers. We also thank the Steering Committee for providing timely advice for running the conference. In particular, we would like to thank Hainan University, China, for providing organizational support and maintaining the conference website, and the City University of Hong Kong, Hong Kong, for hosting the conference website. Finally, we would like to thank the members of the Editorial Board of *Lecture Notes in Computer Science* and the editors at Springer for their encouragement and cooperation throughout the preparation of this conference.

July 2020   Minming Li

Organization

Program Committee

Xiaohui Bei	Nanyang Technological University, Singapore
Xin Chen	Liaoning University of Technology, China
Ran Duan	Tsinghua University, China
Haodi Feng	Shandong University, China
Mordecai J. Golin	The Hong Kong University of Science and Technology, Hong Kong
Siyao Guo	New York University, USA
Xin Han	Dalian University of Technology, China
Wing-Kai Hon	National Tsing Hua University, Taiwan
Naoyuki Kamiyama	Kyushu University, Japan
Yuqing Kong	Peking University, China
Piotr Krysta	The University of Liverpool, UK
Minming Li	City University of Hong Kong, Hong Kong
Shimin Li	Winona State University, USA
Zhengyang Liu	Beijing Institute of Technology, China
Giorgio Lucarelli	LCOMS, University of Lorraine, France
Kim Thang Nguyen	IBISC, University Paris-Saclay, France
Yota Otachi	Nagoya University, Japan
Pan Peng	The University of Sheffield, UK
Richard Peng	Georgia Institute of Technology, USA
Rob van Stee	University of Siegen, Germany
Changjun Wang	Beijing University of Technology, China
Zihe Wang	Shanghai University of Finance and Economics, China
Chenchen Wu	Tianjin University of Technology, China
Deshi Ye	Zhejiang University, China
Cheng Yukun	Zhejiang University of Finance and Economics, China
Chihao Zhang	Shanghai Jiao Tong University, China
Jialin Zhang	Institute of Computing Technology, Chinese Academy of Sciences, China
Peng Zhang	School of Software, Shandong University, China
Yong Zhang	Shenzhen Institutes of Advanced Technology, Chinese Academy of Sciences, China
Yingchao Zhao	Caritas Institute of Higher Education, Hong Kong

Additional Reviewers

Li, Qian
Qiu, Guoliang
Shan, Xiaohan
Xu, Yicheng

Contents

Complexity Results for the Proper Disconnection of Graphs

You Chen, Ping Li, Xueliang Li$^{(\boxtimes)}$, and Yindi Weng

Center for Combinatorics and LPMC, Nankai University, Tianjin 300071, China
chen_you@163.com, wjlpqdxs@163.com, lxl@nankai.edu.cn, 1033174075@qq.com

Abstract. For an edge-colored graph G, a set F of edges of G is called a *proper edge-cut* if F is an edge-cut of G and any pair of adjacent edges in F are assigned by different colors. An edge-colored graph is called *proper disconnected* if for each pair of distinct vertices of G there exists a proper edge-cut separating them. For a connected graph G, the *proper disconnection number* of G, denoted by $pd(G)$, is defined as the minimum number of colors that are needed to make G proper disconnected. In this paper, we first show that it is NP-complete to decide whether a given k-edge-colored graph G with $\Delta(G) = 4$ is proper disconnected. Then, for a graph G with $\Delta(G) \le 3$ we show that $pd(G) \le 2$ and determine the graphs with $pd(G) = 1$ and 2 in polynomial time, respectively, when the set of vertices with degree 3 in G is an independent set. Finally, we show that for a general graph G, deciding whether $pd(G) = 1$ is NP-complete, even if G is bipartite.

Keywords: Edge-coloring · Proper edge-cut · Proper disconnection number · Complexity · NP-complete

1 Introduction

All graphs considered in this paper are finite, simple and undirected. Let $G = (V(G), E(G))$ be a nontrivial connected graph with vertex set $V(G)$ and edge set $E(G)$. For a vertex $v \in V$, the *open neighborhood* of v is the set $N(v) = \{u \in V(G) | uv \in E(G)\}$ and $d(v) = |N(v)|$ is the *degree* of v, and the *closed neighborhood* is the set $N[v] = N(v) \cup \{v\}$. For any notation and terminology not defined here, we follow those used in [4].

For a graph G and a positive integer k, let $c : E(G) \to [k]$ be an edge-coloring of G, where and in what follows $[k]$ denotes the set $\{1, 2, \ldots, k\}$ of integers. For an edge e of G, we denote the color of e by $c(e)$. If adjacent edges of G receive different colors under c, the edge-coloring c is called *proper*. The *chromatic index* of G, denoted by $\chi'(G)$, is the minimum number of colors needed in a proper edge-coloring of G.

In graph theory, paths and cuts are two dual concepts. By Menger's Theorem, paths are in the same position as cuts are in studying graph connectivity.

Supported by NSFC No. 11871034 and 11531011.

M. Li (Ed.): FAW 2020, LNCS 12340, pp. 1–12, 2020.
https://doi.org/10.1007/978-3-030-59901-0_1

Chartrand et al. in [7] introduced the concept of *rainbow connection* of graphs. *Rainbow disconnection*, which is a dual concept of rainbow connection, was introduced by Chartrand et al. [6]. An *edge-cut* of a graph G is a set R of edges such that $G - R$ is disconnected. If any two edges in R have different colors, then R is a *rainbow edge-cut*. An edge-coloring is called a *rainbow disconnection coloring* of G if for every two vertices of G, there exists a rainbow edge-cut in G separating them. For a connected graph G, the *rainbow disconnection number* of G, denoted $rd(G)$, is the smallest number of colors required for a rainbow disconnection coloring of G. A rainbow disconnection coloring using $rd(G)$ colors is called an rd-*coloring* of G. In [2] the authors obtained many results on the rainbow disconnection number.

Andrews et al. [1] and Borozan et al. [5] independently introduced the concept of *proper connection* of graphs. Inspired by the concept of rainbow disconnection and proper connection of graphs, the authors [3] introduced the concept of proper disconnection of graphs. For an edge-colored graph G, a set F of edges of G is a *proper edge-cut* if F is an edge-cut of G and any pair of adjacent edges in F are assigned by different colors. For any two vertices x, y of G, an edge set F is called an x-y proper edge-cut if F is a proper edge-cut and F separates x and y in G. An edge-colored graph is called *proper disconnected* if for each pair of distinct vertices of G there exists a proper edge-cut separating them. For a connected graph G, the *proper disconnection number* of G, denoted by $pd(G)$, is defined as the minimum number of colors that are needed to make G proper disconnected, and such an edge-coloring is called a pd-*coloring*. From [3], we know that if G is a nontrivial connected graph, then $1 \leq pd(G) \leq rd(G) \leq \chi'(G) \leq \Delta(G) + 1$.

Graphs are useful in many real world applications. Colors are used on edges or vertices of graphs to show more information. The proper disconnection of graphs also has applications the real world problems. One of them is stated as follows. In the circulation of goods between cities, we want to intercept some illegal goods, such as drugs and wildlife, and get the interception position. We need to arrange persons to intercept target goods on the roads between cities. When someone intercept successfully, they send out a signal with some frequency to feedback their location. In order to solve this practical problem, we translate it into a coloring problem as follows. Denote each city by a vertex. We assign an edge between two vertices if the corresponding cities are connected directly by a transportation road, and assign a color to each edge based on the frequency used by persons on the corresponding road for communication. Suppose the goods are transported from city A to city B (the corresponding vertices are also denoted by A and B). We only need to intercept on these roads which are corresponding to an A-B edge-cut in G. Preset an A-B edge-cut in G. For the purpose of arranging persons to intercept quickly and intercepting goods successfully, each city whose corresponding vertex is incident with an edge of the A-B edge-cut arranges its own persons to intercept. When the persons on some road intercept the goods successfully, they send out signals to cities which they are from. To feedback the exact location of road on which persons intercept goods, the frequencies used by persons of the same city on different roads should be distinct. Since frequencies are expensive, it is hoped that the number of frequencies is as small as possible. Then the minimum number of frequencies required in this problem is precisely the proper disconnection number of the corresponding graph.

This paper is organized as follows. In Sect. 2, we first show that it is NP-complete to decide whether a given k-edge-colored graph G with $\Delta(G) = 4$ is proper disconnected. Then, for a graph G with $\Delta(G) \leq 3$ we show that $pd(G) \leq 2$, and determine the graphs with $pd(G) = 1$ and 2 in polynomial time, respectively. We further show that if G is a graph with $\Delta(G) = 3$ and the set of vertices with degree 3 in G is an independent set, then deciding whether $pd(G) = 1$ is solvable in polynomial time. Then we propose a question for further study: Is it true that deciding whether $pd(G) = 1$ is solvable in polynomial time for a graph G with $\Delta(G) = 3$? In Sect. 3, we show that for a general graph G, deciding whether $pd(G) = 1$ is NP-complete, even if the graph is bipartite.

2 Hardness Results for Graphs with Maximum Degree Four

In this section, we show that it is NP-complete to decide whether a given k-edge-colored graph G with $\Delta(G) = 4$ is proper disconnected. Then we give the proper disconnection numbers of graphs with $\Delta(G) \leq 3$, and propose a question.

We first give some notations. For an edged-colored graph G, let F be a proper edge-cut of G. If F is a matching, then F is called a *matching cut*. Furthermore, if F is an x-y proper edge-cut for vertices $x, y \in G$, then F is called an x-y *matching cut*. For a vertex v of G, let E_v denote the edge set consisting of all edges incident with v in G.

We can obtain the following results by means of a reduction from the NAE-3-SAT problem. We now present the NAE-3-SAT problem, which is NP-complete; see [8,11].

Problem: Not-All-Equal 3-Sat (NAE-3-SAT)

Instance: A set C of clauses, each containing 3 literals from a set of boolean variables.

Question: Can truth value be assigned to the variables so that each clause contains at least one true literal and at least one false literal?

Given a formula ϕ with variable $x_1, \cdots, x_n$, let $\phi = c_1 \wedge c_2 \wedge \cdots \wedge c_m$, where $c_i = (l_1^i \vee l_2^i \vee l_3^i)$. Then $l_j^i \in \{x_1, \overline{x_1}, \cdots, x_n, \overline{x_n}\}$ for each $i \in [m]$ and $j \in [3]$.

We will construct a graph G_ϕ below. The graphs I_j and C_i are shown in Fig. 1 where $j \in [n]$ and $i \in [m]$. Each graph C_i has two pairs of parallel edges. The graph G_ϕ (see Fig. 2) is obtained from mutually disjoint graphs I_j and C_i by adding a pair of parallel edges between z and w if z, w satisfy one of the following conditions:

1. $z = a_i$ and $w = a_{i+1}$ for some $i \in [n + 2m - 1]$;
2. $z = b_i$ and $w = b_{i+1}$ for some $i \in [n + 2m - 1]$;
3. $z = x_j$, $w = l_t^i$ and $x_j = l_t^i$ for some $j \in [n], t \in [3]$ and $i \in [m]$;
4. $z = \overline{x_j}$, $w = l_t^i$ and $\overline{x_j} = l_t^i$ for some $j \in [n], t \in [3]$ and $i \in [m]$.

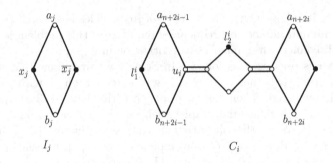

Fig. 1. The graphs I_j and C_i.

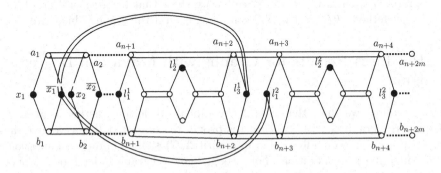

Fig. 2. The graph G_ϕ with $l_3^1 = l_1^2 = \overline{x_1}$.

In fact, the graph G_ϕ was constructed in [10] (in Sect. 3.2). It is obvious that each vertex of G_ϕ with degree greater than four is a vertex with even degree. Moreover, there are two simple edges incident with this kind of vertex, and the other edges incident with the vertex are some pairs of parallel edges. The authors proved that G_ϕ has a matching cut if and only if the corresponding instance ϕ of NAE-3-SAT problem has a solution.

We present a *star structure* as shown in Fig. 3(1). Each vertex z_i is called a *tentacle*. A star structure is a *k-star structure* if it has k tentacles. For a vertex y of G_ϕ with $d_{G_\phi}(y) = 2t + 2 > 4$, assume $N(y) = \{w_1, \cdots, w_{t+2}\}$ such that w_{t+1}, w_{t+2} connect y by a simple edge respectively, and w_i connects y by a pair of parallel edges for $i \in [t]$. Now we define an operation $\mathcal{O}$ on vertex y: replace y by a $(t + 1)$-star structure with tentacles $z_1, \cdots, z_{t+1}$ such that w_i and z_{t+1} for $i \in \{t + 1, t + 2\}$ are connected by a simple edge, and z_i and w_i are connected by parallel edges for $i \in [t]$. As an example, Fig. 3(2) shows the operation $\mathcal{O}$ on vertex y with degree 16. We apply the operation $\mathcal{O}$ on each vertex of degree greater than four, and then subdivide one of each pair of parallel edges using a new vertex in G_ϕ. Denote the resulting graph by G'_ϕ, which is a simple graph. The graph G'_ϕ was also defined in [10], and the authors proved that G'_ϕ has a matching cut if and only if the corresponding instance ϕ of NAE-3-SAT problem has a solution.

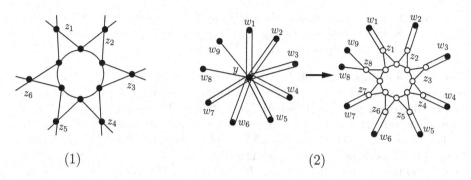

Fig. 3. (1) A 6-star structure with tentacles $z_1, \cdots, z_6$, and (2) the operation $\mathcal{O}$ on vertex y with degree 16.

Now we construct a graph, denoted by H_ϕ, obtained from G_ϕ by operations as follows. Add two new vertices u and v. Connect u and each vertex of $\{a_1, a_{n+2m}\}$ by a pair of parallel edges, and connect v and each vertex of $\{b_1, b_{n+2m}\}$ by a pair of parallel edges. We apply the operation $\mathcal{O}$ on each vertex of degree greater than four in H_ϕ, and then subdivide one of each pair of parallel edges using a new vertex. Denote the resulting graph by H'_ϕ (see Fig. 4), which is a simple graph. Observe that $\Delta(H'_\phi) = 4$. Since a minimal matching cut cannot contain any edge in a triangle, we have that there is a u-v matching cut in H'_ϕ if and only if there is a matching cut in G'_ϕ. Thus, there is a u-v matching cut in H'_ϕ if and only if the instance ϕ of NAE-3-SAT problem has a solution.

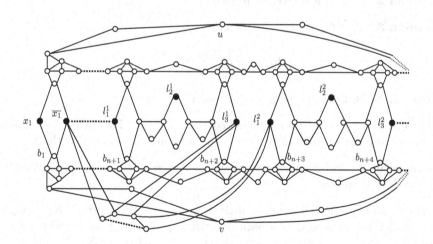

Fig. 4. The graph H'_ϕ with $l_3^1 = l_1^2 = \overline{x_1}$.

Theorem 1. *For a fixed positive integer k, let G be a k-edge-colored graph with $\Delta(G) = 4$, and let u, v be any two specified vertices of G. Then deciding whether there is a u-v proper edge-cut in G is NP-complete.*

Proof. For a connected graph G with an edge-coloring $c : E(G) \to [k]$ and an edge-cut D of G, let $M_i = \{e \mid e \in D$ and $c(e) = i\}$ for $i \in [k]$. Then D is a proper edge-cut if and only if each M_i is a matching. Therefore, deciding whether a given edge-cut of an edge-colored graph is a proper edge-cut is in P.

For an instance ϕ of the NAE-3-SAT problem, we can obtain the corresponding graph H'_ϕ as defined above. Then there is a vertex, say y', of H'_ϕ with degree two. Let G be a graph obtained from H'_ϕ and a path P of order k by identifying y' and one end of P. Then $\Delta(G) = 4$. We color each edge of $G - E(P)$ by 1 and color $k - 1$ edges of P by $2, 3, \cdots, k$, respectively. Then the edge-coloring is a k-edge-coloring of G, and there is a u-v proper edge-cut in G if and only if there is a u-v matching cut in H'_ϕ. Thus, we have that there is a u-v proper edge-cut in G if and only if the instance ϕ of NAE-3-SAT problem has a solution.

Corollary 1. *For a fixed positive integer k, let G be a k-edge-colored graph with $\Delta(G) = 4$. Then it is NP-complete to decide whether G is proper disconnected.*

Even though it is still not clear for the computational complexity of deciding whether a graph with maximum degree at most three is proper disconnected, we will show that $pd(G) \le 2$ for a graph G with $\Delta(G) \le 3$ and then determine the graphs with $pd(G) = 1$ and 2, respectively. Some useful results are given as follows, which will be used in the sequel.

Theorem 2. [3] *If G is a tree, then $pd(G) = 1$.*

Theorem 3. [3] *If C_n be a cycle, then*

$$\mathrm{pd}(C_n) = \begin{cases} 2, \ if \ n = 3, \\ 1, \ if \ n \ge 4. \end{cases}$$

Theorem 4. [3] *For any integer $n \ge 2$, $pd(K_n) = \lceil \frac{n}{2} \rceil$.*

Theorem 5. [3] *Let G be a nontrivial connected graph. Then $pd(G) = 1$ if and only if for any two vertices of G, there is a matching cut separating them.*

Theorem 6. [4] *(Petersen's Theorem) Every 3-regular graph without cut edges has a perfect matching.*

For a simple connected graph G, if $\Delta(G) = 1$, then G is the graph K_2, a single edge. If $\Delta(G) = 2$, then G is a path of order $n \ge 3$ or a cycle. By Theorems 2 and 3, for a connected graph G with $\Delta(G) \le 2$, we have $pd(G) = 1$ if and only if G is a path or a cycle of order $n \ge 4$, and $pd(G) = 2$ if and only if G is a triangle.

Next, we will present the proper disconnection numbers of graphs with maximum degree 3. At first, we give the proper disconnection numbers of 3-regular graphs.

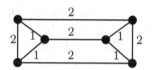

Fig. 5. The graph G_0

Lemma 1. *If G is a 3-regular connected graph without cut edges, then $pd(G) \leq$ 2.*

Proof. Let G_0 be a graph by connecting two triangles with 3 matching edges, and we color G_0 with two colors as shown in Fig. 5. Obviously, it is a proper disconnection coloring of G_0. Now we consider 2-edge-connected 3-regular graphs G except G_0. By Theorem 6, there exists a perfect matching M in G. We define an edge-coloring c of G as follows. Let $c(M) = 2$. If $E(G) \backslash M$ contains triangles, then we color one edge of each triangle by color 2. We color the remaining edges by color 1. Since $G \backslash M$ is the union of some disjoint cycles, we denote these disjoint cycles by $C_1, C_2, \cdots C_t$. Let x, y be two vertices of G. If x and y belong to different cycles of $C_1, C_2, \cdots C_t$, then M is an x-y proper edge-cut. If x and y belong to the same cycle $C_i(i \in [t])$, then there are two cases to discuss.

Case 1. $|C_i| \geq 4$.

Since $|C_i| \geq 4$, there exist two x-y paths P_1, P_2 in C_i. We choose two non-adjacent edges e_1, e_2 respectively from P_1, P_2. Then $M \cup \{e_1, e_2\}$ is an edge-cut between x and y. Since $c(M) = 2$ and $c(e_1) = c(e_2) = 1$, $M \cup \{e_1, e_2\}$ is an x-y proper edge-cut.

Case 2. $|C_i| = 3$.

Since $x, y \in C_i$, we can assume $C_i = xyz$. Let $N(x) = \{y, z, x_0\}$ and $N(x_0) = \{x, x_1, x_2\}$. Assume $x_0 \in C_k, k \in [t] \backslash \{i\}$.

Subcase 2.1. $c(xy) = 1$.

Assume $c(yz) = 1$ and $c(xz) = 2$. We note that $x_1 \notin N(z)$ or $x_2 \notin N(z)$. Without loss of generality, $x_2 \notin N(z)$.

For $|C_k| \geq 4$, we have $c(x_0 x_2) = 1$. Then $E_{x_2} \backslash \{x_0 x_2\}$ have different colors. So, $\{xy, xz, x_0 x_1\} \cup E_{x_2} \backslash \{x_0 x_2\}$ is an x-y proper edge-cut.

For $|C_k| = 3$, if $c(x_0 x_1) \neq c(x_0 x_2)$, we get that $\{xy, xz, x_0 x_1, x_0 x_2\}$ is an x-y proper edge-cut. Now consider $c(x_0 x_1) = c(x_0 x_2) = 1$. If $x_1 \in N(z)$, then $c(x_1 z) = 2$. Since $G \neq G_0$, we have $x_2 \notin N(y) \cup N(z)$. So $(E_y \backslash \{yz\}) \cup \{xz, x_0 x_1, x_1 x_2\}$ is an x-y proper edge-cut. If $x_1 \notin N(z)$, then we denote $E_{x_1} \backslash \{x_0 x_1, x_1 x_2\}$ by e_1 and denote $E_{x_2} \backslash \{x_0 x_2, x_1 x_2\}$ by e_2. It is clear that $e_1, e_2 \in M$. So, $c(e_1) = c(e_2) = 2$. We get that $\{xy, xz, e_1, e_2\}$ is an x-y proper edge-cut.

Subcase 2.2. $c(xy) = 2$.

In Subcase 2.1, if $c(xy) = 1$, then the x-y proper edge-cut is also an x-z proper edge-cut. So, we have proved Subcase 2.2.

Let $H(v)$ be a connected graph with one vertex v of degree two and the remaining vertices of degree three. We assume that the neighbors of v in $H(v)$

are v_1 and v_2, respectively. If v_1, v_2 are adjacent, then we denote it by $H_1(v)$. Otherwise, we denote it by $H_2(v)$. Let $H_1'(v)$ be the graph obtained by replacing the vertex v by a diamond. Let $H_2'(v)$ be the graph by replacing the path v_1vv_2 of $H_2(v)$ by an new edge v_1v_2. See Fig. 6.

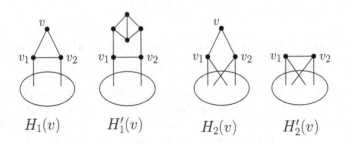

$$H_1(v) \qquad H_1'(v) \qquad H_2(v) \qquad H_2'(v)$$

Fig. 6. The graph process

Lemma 2. *If G is a 3-regular graph of order n $(n \geq 4)$, then $pd(G) \leq 2$.*

Proof. We proceed by induction on the order n of G. A 3-regular graph of order 4 is K_4 and $pd(K_4) = 2$ from Theorem 4. So, the result is true for $n = 4$. Suppose that if H is a 3-regular graph of order n $(n \geq 4)$, then $pd(G) \leq 2$. Let G be a 3-regular graph of order $n + 1$. We will show $pd(G) \leq 2$. If G has no cut edge, then $pd(G) \leq 2$ from Lemma 1. So, we consider G having a cut edge, say uv $(u, v \in V(G))$. We delete the cut edge uv, then there are two components containing u and v, respectively, say G_1, G_2. Since G is 3-regular, we have $|V(G_1)| \geq 5$ and $|V(G_2)| \geq 5$. Thus, $5 \leq |V(G_1)| \leq n - 4$ and $5 \leq |V(G_2)| \leq n - 4$. Obviously, G_1 and G_2 are the graphs $H(u)$, $H(v)$, respectively. We first show the following claims.

Claim. $pd(H_1(u)) \leq 2$.

Proof. Let u_1 and u_2 be two neighbors of u in $H_1(u)$. Assume that the neighbors of u_1 and u_2 in $H_1(u)$ are $\{u, u_2, w_1\}$, $\{u, u_1, w_2\}$, respectively. The edges u_1w_1, u_1u_2 and u_2w_2 are denoted by e_1, e_2, e_3. Let $A = \{u, u_1, u_2\}$ and $B = V(H_1(u))\backslash A$. Since $|V(G_1)| \leq n - 4$, we have $|V(H_1'(u))| \leq n - 1$. Obviously, $H_1'(u)$ is 3-regular. Then $pd(H_1'(u)) \leq 2$ by the induction. Let c' be a proper disconnection coloring of $H_1'(u)$ with two colors. For any two vertices p and q of $H_1'(u)$, let R_{pq} be a p-q proper edge-cut of $H_1'(u)$. There are two cases to discuss.

Case 1. $c'(e_1) = c'(e_2)$ or $c'(e_2) = c'(e_3)$.

Without loss of generality, we assume $c'(e_1) = c'(e_2) = 1$. We define an edge-coloring c of $H_1(u)$ as follows. Let $c(uu_1) = 2$, $c(uu_2) = 1$ and $c(e) = c'(e)$ $(e \in E(H_1(u))\backslash\{uu_1, uu_2\})$. Let x and y be two vertices of $H_1(u)$. If they are both in $H_1(u)\backslash\{u\}$, then $(R_{xy} \cap E(H_1(u))) \cup \{uu_1\}$ is an x-y proper edge-cut of

$H_1(u)$. If $x = u$ or $y = u$, then E_u is an x-y proper edge-cut of $H_1(u)$. So, c is a proper disconnection coloring of $H_1(u)$. Thus, $pd(H_1(u)) \leq 2$.

Case 2. $c'(e_1) = c'(e_3) \neq c'(e_2)$.

Assume $c'(e_1) = c'(e_3) = 1$ and $c'(e_2) = 2$. Define an edge-coloring c of $H_1(u)$ as follows. Let $c(uu_1) = 2$, $c(uu_2) = 1$ and $c(e) = c'(e)$ ($e \in E(H_1(u)) \backslash \{uu_1, uu_2\}$). Let x and y be two vertices of $H_1(u)$. If $x = u$ or $y = u$, then E_u is an x-y proper edge-cut of $H_1(u)$. If $x, y \in A \backslash \{u\}$, then $\{uu_2, e_1, e_2\}$ is an x-y proper edge-cut of $H_1(u)$. If $x \in A \backslash \{u\}, y \in B$ or $x \in B, y \in A \backslash \{u\}$, then $\{e_1, e_3\}$ is an x-y proper edge-cut of $H_1(u)$. Considering $x, y \in B$, if $e_1, e_3 \notin R_{xy}$, then $(R_{xy} \cap E(H_1(u))) \cup \{uu_2\}$ is an x-y proper edge-cut of $H_1(u)$. Otherwise, i.e., $e_1 \in R_{xy}$ or $e_3 \in R_{xy}$, then $R_{xy} \cap E(H_1(u))$ is an x-y proper edge-cut of $H_1(u)$. So, c is a proper disconnection coloring of $H_1(u)$. Thus, $pd(H_1(u)) \leq 2$.

Claim. $pd(H_2(u)) \leq 2$.

Proof. Assume that the neighbors of u in $H_2(u)$ are u_1 and u_2. Since $|V(H'_2(u))| < |V(G_2)|$ and $H'_2(u)$ is 3-regular. Then $pd(H'_2(u)) \leq 2$ by the induction. Let c' be a proper disconnection coloring of $H'_2(u)$ with two colors. We define an edge-coloring c of $H_2(u)$ as follows: $c(uu_1) = 1, c(uu_2) = 2$ and $c(e) = c'(e)$ ($e \in E(H_2(u)) \backslash \{uu_1, uu_2\}$). Assume $c'(u_1u_2) = c(uu_i)$ ($i = 1$ or 2). Then for any two vertices x and y of $H_2(u)$, if $x = u$ or $y = u$, then E_u forms an x-y proper edge-cut. Otherwise, assume the x-y proper edge-cut in $H'_2(u)$ is R. If $u_1u_2 \notin R$, then R is an x-y proper edge-cut. If $u_1u_2 \in R$, then $(R \cup \{uu_i\}) \backslash \{u_1u_2\}$ is an x-y proper edge-cut. So, c is a proper disconnection coloring of $H_2(u)$. Thus, $pd(H_2(u)) \leq 2$.

So, from above claims we have $pd(G_1) \leq 2$. Similarly, we have $pd(G_2) \leq 2$. Then, there exists a proper disconnection coloring c_0 of $G_1 \cup G_2$ with two colors. Now we assign the color 1 to the cut edge uv. It is a proper disconnection coloring of G. So, $pd(G) \leq 2$.

A *block* of a graph G is a maximal connected subgraph of G that has no cut vertex. Let $\{B_1, B_2, \ldots, B_t\}$ be the set of blocks of G.

Lemma 3. [3] Let G be a nontrivial connected graph. Then $pd(G) = \max\{pd(B_i) \mid i = 1, 2, \ldots, t\}$.

Theorem 7. *If G is a graph of order n with $\Delta(G) = 3$, then $pd(G) \leq 2$. Particularly, if G satisfies the condition of Theorem 5, then $pd(G) = 1$; otherwise, $pd(G) = 2$.*

Proof. If G is a tree, then $pd(G) = 1$ by Theorem 2. Suppose G is not a tree. Let H be a graph as shown in Fig. 7. Call v the *key vertex* of H. Suppose G is a graph with maximum degree three. Let G' be a graph obtained from G by deleting pendent edges one by one. Then $\Delta(G') \leq 3$ and $pd(G) = pd(G')$ by Lemma 3. Let $\{u_1, \cdots, u_t\}$ be the set of 2-degree vertices in G' and $H_1, \cdots, H_t$ be t copies of H such that the key vertex of H_i is v_i ($i \in [t]$). We construct a new graph G'' obtained by connecting v_i and u_i for each $i \in [t]$. Then G'' is a 3-regular graph. By Lemma 2, $pd(G'') \leq 2$. Since G' is a subgraph of G'', then $pd(G') \leq 2$.

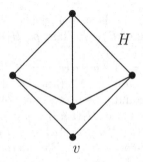

Fig. 7. The graph H.

Theorem 8. *Let G be a connected graph with $\Delta = 3$ such that the set of vertices with degree 3 in G forms an independent set. If G contains a triangle or $K_{2,3}$, then $pd(G) = 2$; otherwise, $pd(G) = 1$.*

Proof. If G contains a triangle or a $K_{2,3}$, then there exist two vertices such that no matching cut separates them. So, $pd(G) = 2$ by Theorem 7. Now consider that G is triangle-free and $K_{2,3}$-free. We proceed by induction on the order n of G. Since $\Delta(G) = 3$, we have $n \geq 4$. If $n = 4$, then the graph G is $K_{1,3}$ and $pd(G) = 1$ by Theorem 2. The result holds for $n = 4$. Assume $pd(G) = 1$ for triangle-free and $K_{2,3}$-free graphs with order n satisfying the condition. Now, consider a graph G with order $n + 1$. Let x and y be two vertices of G.

For $d(x) = 1$, the edge E_x is an x-y matching cut.

For $d(x) = d(y) = 2$, if x and y are adjacent, let x_1, y_1 be another neighbor of x and y, respectively. Let $G' = G - xy$. Then by the induction, there exist an x_1-y_1 matching cut R in G'. Thus, $R \cup \{xy\}$ is an x-y matching cut in G. If x and y are nonadjacent, then assume $N(x) = \{x_1, x_2\}$. Since G contains no triangles, then x_1 and x_2 are nonadjacent. There are two cases to consider. If $d(x_1) = 2$, then let u_1 be another neighbor of x_1, and then $\{xx_2, x_1u_1\}$ is an x-y matching cut. If $d(x_1) = d(x_2) = 3$, let $N(x_1) = \{x, u_1, u_2\}$ and $N(x_2) = \{x, v_1, v_2\}$. There are two cases to consider. If $\{u_1, u_2\} \cap \{v_1, v_2\} \neq \emptyset$, assume $u_1 = v_1$. Let w, q be another neighbor of u_2 and v_2, respectively. If $y \neq v_2$, then $\{xx_1, x_2u_1, v_2q\}$ is an x-y matching cut. Otherwise, $\{xx_2, x_1u_1, u_2w\}$ is an x-y matching cut. Assume $\{u_1, u_2\} \cap \{v_1, v_2\} = \emptyset$. Let w, q be another neighbor of u_1 and u_2, respectively. If $y = u_1$, then $\{xx_2, x_1u_1, u_2q\}$ is an x-y matching cut. Otherwise, $\{xx_2, x_1u_2, u_1w\}$ is an x-y matching cut.

For $d(x) = 3$ (or $d(y) = 3$), assume $N(x) = \{x_1, x_2, x_3\}$. Since the set of vertices with degree 3 in G forms an independent set, the neighbors of x have degree at most two. Since G is $K_{2,3}$-free, there exists at least one vertex in $N(x)$ which has only one common neighbor x with the others in $N(x)$. Without loss of generality, say x_1. Let $N(x_1) = \{x, s_1\}, N(x_2) = \{x, s_2\}$ and $N(x_3) = \{x, s_3\}$ ($s_2 = s_3$ is possible). If x and y are not adjacent, then $\{x_1s_1, x_2s_2, xx_3\}$ is an x-y matching cut. If x and y are adjacent, there are three cases to consider. When $y = x_2$ (or x_3), we have $\{x_1s_1, xy, x_3s_3\}$ (or $\{x_1s_1, xy, x_2s_2\}$) is an x-y matching cut. When $y = x_1$ and $s_2 = s_3$, if $d(s_2) = 2$, then $\{xy\}$ is an x-y matching cut; if $d(s_2) = 3$, then assume $N(s_2) = \{x_2, x_3, p_1\}$, and then $\{xy, s_2p_1\}$ is an x-y matching cut. When $y = x_1$ and $s_2 \neq s_3$, we have $\{xy, x_2s_2, x_3s_3\}$ is an x-y matching cut. Thus, $pd(G) = 1$ by Theorem 5.

Corollary 2. *Let G be a connected graph with $\Delta = 3$. If the set of vertices with degree 3 in G forms an independent set, then deciding whether $pd(G) = 1$ is solvable in polynomial time.*

We naturally put forward the following question.

Question 1. *Let G be a connected graph with $\Delta = 3$. Is it true that deciding whether $pd(G) = 1$ is solvable in polynomial time?*

3   Hardness Results for Bipartite Graphs

In fact, by Corollary 1, we know that it is NP-complete to decide whether $pd(G) = 1$ for a general graph G. In this section, we will further show that given a bipartite graph G, deciding whether $pd(G) = 1$ is NP-complete.

Let G be a simple connected graph. We employ the idea used in [9] to construct a new graph G^*. Let G^* be a graph obtained from G by replacing each edge by a 4-cycle. Then G^* has two types of vertices: *old* vertices, which are vertices of G, and *new* vertices, which are not vertices of G. For example, for an edge $e = uv \in E(G)$, replace it by a 4-cycle $C_e = uxvyu$. Then u, v are old vertices and x, y are new vertices. Observe that all new vertices of G^* have degree two, and each edge of G^* connects an old vertex to a new vertex. Clearly, G^* is a bipartite graph with one side of bipartition consisting only of vertices of degree 2.

Lemma 4. *Let G be a simple connected graph. Then $pd(G) = 1$ if and only if $pd(G^*) = 1$.*

Proof. Suppose $pd(G^*) = 1$. For any two vertices x, y of G, x, y are old vertices in G^*. By Theorem 5, there exists an x-y matching cut F in G^*. Then F consists of pairs of matching edges in the same 4-cycle. Let F' be the edge set obtained by replacing each pair of matching edges of F in the same 4-cycle by the edge to which the 4-cycle corresponds in G. Then F' is an x-y matching cut in G.

Suppose $pd(G) = 1$. Then for any two vertices u, v of G, there is a u-v matching cut. We denote it by F_{uv}. Let F be an edge subset of $E(G)$. Choose two matching edges from each 4-cycle to which each edge of F corresponds in G^*. Denote the edge set by F^*.

For any two vertices x, y in G^*, if x, y are old vertices, then F^*_{xy} is an x-y matching cut in G^*. If x is an old vertex and y is a new vertex, there are two cases to consider. If x, y are in the same 4-cycle, assume that the 4-cycle is $xyzwx$. We know $xz \in F_{xz}$ in G. Then $(F_{xz} \setminus \{xz\})^* \cup \{xy, zw\}$ is an x-y matching cut in G^*. If x, y are in different 4-cycles, say $C_1 = xv_1uv_2x$ and $C_2 = yz_1wz_2y$. If there are no x-z_2 paths in $G - F_{xz_1}$, then $F^*_{xz_1}$ is an x-y matching cut in G^*.

Otherwise, we know $z_1 z_2 \in F_{xz_1}$. Then $(F_{xz_1} \backslash \{z_1 z_2\})^* \cup \{yz_2, wz_1\}$ is an x-y matching cut in G^*. If x, y are new vertices, there are two cases to consider. If x, y are in the same 4-cycle, assume that the 4-cycle is $\{uxvyu\}$. Then F_{uv}^* is an x-y matching cut in G^*. If x, y are in different 4-cycles, say $C_1 = u_1 x u_2 v u_1$ and $C_2 = z_1 y z_2 w z_1$. Denote the component containing u_1 by C and the remaining part by $\bar{C}$ in $G - F_{u_1 u_2}$. If $\{z_1 z_2\} \subseteq \bar{C}$, then $(F_{u_1 u_2} \backslash \{u_1 u_2\})^* \cup \{u_1 v, u_2 x\}$ is an x-y matching cut in G^*. If $z_1 \in C$ and $z_2 \in \bar{C}$, then $z_1 z_2 \in F_{u_1 u_2}$. Then, $(F_{u_1 u_2} \backslash \{u_1 u_2, z_1 z_2\})^* \cup \{u_1 v, u_2 x, z_1 y, z_2 w\}$ is an x-y matching cut in G^*. If $\{z_1 z_2\} \subseteq C$, then $(F_{u_1 u_2} \backslash \{u_1 u_2\})^* \cup \{u_1 x, u_2 v\}$ is an x-y matching cut in G^*.

From the above Lemma 4, we can immediately get the following result.

Theorem 9. *Given a bipartite graph G, deciding whether $pd(G) = 1$ is NP-complete.*

Acknowledgement. The authors would like to thank the reviewers and editor for their helpful suggestions and comments.

References

1. Andrews, E., Laforge, E., Lumduanhom, C., Zhang, P.: On proper-path colorings in graphs. J. Combin. Math. Combin. Comput. **97**, 189–207 (2016)
2. Bai, X., Chang, R., Li, X.: More on rainbow disconnection in graphs. Discuss. Math. Graph Theory (in press). https://doi.org/10.7151/dmgt.2333
3. Bai, X., Chen, Y., Ji, M., Li, X., Weng, Y., Wu, W.: Proper disconnection in graphs. arXiv:1906.01832 [math.CO]
4. Bondy, J.A., Murty, U.S.R.: Graph Theory. Springer, London (2008)
5. Borozan, V., et al.: Proper connection of graphs. Discret. Math. **312**, 2550–2560 (2012)
6. Chartrand, G., Devereaux, S., Haynes, T.W., Hedetniemi, S.T., Zhang, P.: Rainbow disconnection in graphs. Discuss. Math. Graph Theor. **38**, 1007–1021 (2018)
7. Chartrand, G., Johns, G.L., McKeon, K.A., Zhang, P.: Rainbow connection in graphs. Math. Bohem. **133**, 85–98 (2008)
8. Darmann, A., Döcker, J.: On simplified NP-complete variants of Not-All-Equal 3-SAT and 3-SAT. arXiv:1908.04198 [cs.CC]
9. Moshi, A.M.: Matching cutsets in graphs. J. Graph Theory **13**, 527–536 (1989)
10. Patrignani, M., Pizzonia, M.: The complexity of the matching-cut problem. In: Brandstädt, A., Le, V.B. (eds.) WG 2001. LNCS, vol. 2204, pp. 284–295. Springer, Heidelberg (2001). https://doi.org/10.1007/3-540-45477-2_26
11. Schaefer, T.J.: The complexity of satisfiability problems. In: Procceedings of the 10th Annual ACM Symposium on Theory of Computing, pp. 216–226. ACM, New York (1978). https://doi.org/10.1145/800133.804350

A Polynomial Delay Algorithm for Enumerating 2-Edge-Connected Induced Subgraphs

Yusuke Sano, Katsuhisa Yamanaka(✉) , and Takashi Hirayama

Faculty of Science and Engineering, Iwate University, Ueda 4-3-5, Morioka, Iwate, Japan
hasuko@kono.cis.iwate-u.ac.jp, yamanaka@cis.iwate-u.ac.jp,
hirayama@kono.cis.iwate-u.ac.jp

Abstract. The problem of enumerating connected induced subgraphs of a given graph is classical and studied well. It is known that connected induced subgraphs can be enumerated in constant time for each subgraph. In this paper, we focus on highly connected induced subgraphs. The most major concept of connectivity on graphs is vertex connectivity. For vertex connectivity, some enumeration problem settings are proposed and enumeration algorithms are proposed, such as k-vertex connected spanning subgraphs. In this paper, we focus on another major concept of graph connectivity, edge-connectivity. This is motivated by the problem of finding evacuation routes in road networks. In evacuation routes, edge-connectivity is important, since highly edge-connected subgraphs ensure multiple routes between two vertices. In this paper, we consider the problem of enumerating 2-edge-connected induced subgraphs of a given graph. We present an algorithm that enumerates 2-edge-connected induced subgraphs of an input graph G with n vertices and m edges. Our algorithm enumerates all the 2-edge-connected induced subgraphs in $O(n^3 m |S_G|)$ time, where S_G is the set of the 2-edge-connected induced subgraphs of G. Moreover, by slightly modifying the algorithm, we have a polynomial delay enumeration algorithm for 2-edge-connected induced subgraphs.

Keywords: Enumeration algorithm · 2-edge-connected induced subgraph · Reverse search · Polynomial delay

1 Introduction

Enumerating substructures of enormous data is a fundamental and important problem. An enumeration is one of the strong and appealing strategies to discover some knowledge from enormous data in various research areas such as data mining, bioinformatics and artificial intelligence. From this viewpoint, various enumeration algorithms have been designed.

Graphs are used to represent the relationship of objects. In web-graphs, web pages are represented by vertices of graphs and links between web pages are represented by edges. For social network, users are represented by vertices of graphs and their

This work was supported by JSPS KAKENHI Grant Numbers JP18H04091 and JP19K11812.

friendship relations are represented by edges. In the area of bioinformatics, molecular interactions are represented by graphs. To discover valuable knowledge from practical graphs, enumeration algorithms for subgraphs with some properties are studied, such as simple/induced paths [3,8,16,20], simple/induced cycles [3,8,16,20], subtrees [22], spanning trees [16,17,19], k-vertex-connected spanning subgraphs [4,10], k-edge-connected spanning subgraphs [24], maximal k-edge-connected subgraphs [1], cliques [7,12], pseudo cliques [18], k-degenerate subgraphs [5], matchings [19], induced matchings [11], connected induced subgraphs [2,14,19], and so on. Several years ago, a good textbook on enumeration has been published [13]. Very recently, Conte and Uno [6] proposed a framework for enumerating maximal subgraphs with various properties in polynomial delay.

Some of existing results above focus on closely related subgraphs. This comes from the fact that some applications on knowledge discovery need to find closely related community on graph structures. In this paper, we focus on highly edge-connected induced subgraphs. This is motivated by the problem of finding evacuation routes of road networks in time of disaster. In the time of disaster, it is easy to imagine that many roads are broken. In the situation that we know only one route from the current position to a shelter, nobody can ensure that the route can be passed through in safety. From this point of view, the problem of finding subgraphs with high edge-connectivity is important, since high edge-connectivity of graphs ensure multiple routes between two places. Now, we have the following question: Can we efficiently enumerate all k-edge-connected induced subgraphs? Here, an efficient enumeration implies an output polynomial or a polynomial delay enumeration.

As the first step toward the question, we focus on 2-edge-connected induced subgraphs as highly edge-connected induced subgraphs. In this paper, we propose an algorithm that enumerates all 2-edge-connected induced subgraphs of a given graph. The algorithm is based on reverse search [2]. First, we define a tree structure, called a family tree, on a set of 2-edge-connected induced subgraphs of a given graph. Then, by traversing the tree, we enumerate all the 2-edge-connected induced subgraphs. For an input graph G with n vertices and m edges, our algorithm runs in $O(n^3 m |S_G|)$, where S_G is the set of the 2-edge-connected induced subgraphs. By applying the alternative output technique by Nakano and Uno [15], we have an enumeration algorithm that runs in polynomial delay.

Due to space limitation, some proofs are omitted.

2   Preliminary

2.1   Graphs and Notations

In this paper, we assume that all graphs are simple, undirected, and unweighted. Let $G = (V(G), E(G))$ be a graph with vertex set $V(G)$ and edge set $E(G)$. We define $n = |V(G)|$ and $m = |E(G)|$. The *neighbor set* of a vertex v, denoted by $N(v)$, is the set of vertices adjacent to v. The *degree* of v, denoted by $d(v)$, is the number of vertices in $N(v)$. A *subgraph* of a graph G is a graph $H = (V(H), E(H))$ such that $V(H) \subseteq V(G)$ and $E(H) \subseteq \{\{u, v\} \mid u, v \in V(H) \text{ and } \{u, v\} \in E(G)\}$.

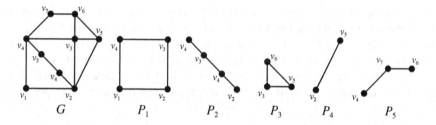

Fig. 1. An example of a closed-ear decomposition. The graph G has a closed-ear decomposition P_1, P_2, P_3, P_4, P_5.

A *path* of G is an alternating sequence $\langle v_1, e_1, v_2, e_2, \ldots, e_{k-1}, v_k \rangle$ of vertices and edges, where $e_i = \{v_i, v_{i+1}\}$ for $1 \leq i \leq k - 1$, such that $e_i \in E(G)$ holds. The *length*, denoted by $|P|$, of a path P is the number of the edges in the path. The path is *simple* if the path contains distinct vertices and distinct edges. Let $P = \langle v_1, e_1, v_2, e_2, \ldots, e_{k-1}, v_k \rangle$ be a simple path. We write $P = \langle v_1, v_2, \ldots, v_k \rangle$ by omitting the internal edges of P. A simple path P is an *open ear* of G if $d(v) = 2$ for each internal vertex v and $d(u) > 2$ for each endpoint u holds. A path $P = \langle v_1, v_2, \ldots, v_k \rangle$ is a *cycle* if $v_1 = v_k$ holds. A cycle is *simple* if a cycle has distinct internal vertices and distinct edges. A simple cycle P is a *closed ear* of G if $d(v_i) = 2$ for $i = 2, 3, \ldots, k - 1$ and $d(v_i) > 2$ for $i = 1$ (or $i = k$).

Let $G_1 = (V(G_1), E(G_1))$ and $G_2 = (V(G_2), E(G_2))$ be two graphs. The union of G_1 and G_2 is the graph $G_1 \cup G_2 = (V(G_1) \cup V(G_2), E(G_1) \cup E(G_2))$. A *decomposition* of a graph G is a list $H_1, H_2, \ldots, H_k$ of subgraphs such that each edge appears in exactly one subgraph in the list and $G = H_1 \cup H_2 \cup \cdots \cup H_k$ holds. A *closed-ear decomposition* of G is a decomposition $P_1, P_2, \ldots, P_k$ such that P_1 is a cycle and P_i for $i \geq 2$ is either an open ear or a closed ear in $P_1 \cup P_2 \cup \cdots \cup P_i$.[1] See Fig. 1 for an example.

An *edge-cut* of G is a set $F \subseteq E(G)$ if the removal of edges in F makes G unconnected. A graph is *k-edge-connected* if every edge-cut has at least k edges. A *bridge* is an edge-cut consisting of one edge. For bridges we have the following characterization:

Theorem 1 ([23], **p. 23**). *An edge is a bridge if and only if it belongs to no cycle.*

From the above theorem, if there is a cycle including an edge e, e is not a bridge. From the definition, we have the following observation.

Observation 1. *A graph is 2-edge-connected if and only if the graph has no bridge.*

A 2-edge-connected graph has another characterization:

Theorem 2 ([23], **p. 164**). *A graph is 2-edge-connected if and only if it has a closed-ear decomposition and every cycle in a 2-edge-connected graph is the initial cycle in some such decomposition.*

An *induced subgraph* of G is a subgraph $H = (V(H), E(H))$ such that $V(H) \subseteq V(G)$ and $E(H) = \{\{u, v\} \mid u, v \in V(H) \text{ and } \{u, v\} \in E(G)\}$. We say that H is a subgraph

---

[1] A closed-ear decomposition is an *ear decomposition* if every ear in the decomposition is an open ear.

of G induced by $V(H)$ and denoted by $G[V(H)]$. Let S be a subset of $V(G)$. We define $H+S$ as the subgraphs induced by $V(H) \cup S$. Similarly, we define $H-S$ as the subgraph induced by $V(H) \backslash S$. Let $H_1 = (V(H_1), E(H_1))$ and $H_2 = (V(H_2), E(H_2))$ be two induced subgraphs of G. We define $H_1 + H_2$ as the subgraphs induced by $V(H_1) \cup V(H_2)$. We define $H_1 - H_2$ as the subgraphs induced by $V(H_1) \backslash V(H_2)$. An induced subgraph is an *induced path* and *induced cycle* if it forms a simple path and simple cycle, respectively.

Observation 2. *Let $G = (V(G), E(G))$ be a 2-edge-connected graph. Then, G has a closed-ear decomposition that ends up with an induced cycle of G.*

Proof. Let C be a induced cycle of G. Since G is 2-edge-connected, from Theorem 2, there exists a closed-ear decomposition that ends up with C. Therefore, the claim is proved.

Now, let $H = (V(H), E(H))$ be a 2-edge-connected induced subgraph of G, and let $S \subseteq V(H)$ be a subset of $V(H)$. A vertex v in S is a *boundary* if v is adjacent to a vertex in $V(H) \backslash S$. A vertex subset S is *removable* if $H - S$ is 2-edge-connected.

Lemma 1. *Let G be a graph, and let H be a 2-edge-connected induced subgraph of G. Suppose that H is not an induced cycle. Then, H has a removable set.*

Proof. From Observation 2, H has a closed decomposition that ends up with an induced cycle. Let $C = (V(C), E(C))$ be such an induced cycle of G. Then, we can observe that $V(H) \backslash V(C)$ is a removable set. □

A removable set S of H is *minimal* if any $S' \subset S$ is not a removable set of H. We have the following properties of minimal removable sets.

Lemma 2. *Let G be a graph, and let H be a 2-edge-connected induced subgraph of G. Let S be a minimal removable set of H. Then, $G[S]$ is connected.*

Proof. Suppose for a contradiction that $G[S]$ is unconnected. Let S' be a subset of S such that $G[S']$ is a connected component in $G[S]$. Then, S' is a removable set, which contradicts to the minimality of S. □

Lemma 3. *Let G be a graph, and let H be a 2-edge-connected induced subgraph of G. Any minimal removable set S, $|S| \geq 2$, of H has exactly two boundaries.*

Proof. Omitted. □

Now, we have the following key lemma.

Lemma 4. *Let G be a graph, and let H be a 2-edge-connected induced subgraph of G. Let S, $|S| \geq 2$, be a minimal removable set of H. Then, $G[S]$ is a path of length $|S| - 1$.*

Proof. Omitted. □

From Lemma 3 and Lemma 4, we can write a minimal removable set as a sequence $S = \langle u_1, u_2, \ldots, u_k \rangle$. Moreover, any internal vertex is not boundary of S. That is, the endpoints of a path are boundaries. Otherwise, since $G[S]$ is a path, an endpoint has degree 1 in H, which contradict to the 2-edge-connectivity of H. From now on, we assume that the two endpoints u_1 and u_k are boundaries of S.

A path $P = \langle w_1, w_2, \ldots, w_\ell \rangle$ with $|V(P)| \geq 2$ of H is an *internal ear* if (1) w_1 and w_k are the two boundaries of $V(P)$ and (2) $d(w_i) = 2$ in H holds for $i = 1, 2, \ldots, \ell$. An internal ear P is *maximal* if there is no internal ear P' such that P' includes P as its subpath. We have the following observation on forms of minimal removable sets.

Observation 3. *Let G be a graph, and let H be a 2-edge-connected induced subgraph of G. Let $S = \langle u_1, u_2, \ldots, u_k \rangle$ be a minimal removable set of H with two boundaries u_1, u_k. Then,*

1. *if $|S| = 1$, S forms a path in H with length 0 (in this case, $u_1 = u_k$) and*
2. *if $|S| \geq 2$, S forms a maximal internal ear of H.*

Proof. Omitted.   □

From Observation 3, a minimal removable set of a 2-edge-connected induced subgraph H forms a maximal internal ear of H. However, note that the reverse direction is not always true.

2.2 Enumeration Algorithms

For algorithms of normal decision problem or optimization problems, we estimate the running time of the whole algorithm as a function of input size. On the other hands, in enumeration problems, we sometimes have exponential outputs for input-size. For enumeration algorithms, we use particular running-time analysis. In this subsection, we introduce some analysis ways for enumeration algorithms.

Let $\mathcal{A}$ be an enumeration algorithm for an enumeration problem Π with input size n and output size α. The algorithm $\mathcal{A}$ is *output polynomial* if $\mathcal{A}$ solves Π in $O(n^c \alpha^d)$ time, where c, d are some constants. The algorithm $\mathcal{A}$ *P-enumerates* if $\mathcal{A}$ solves Π in $O(n^c \alpha)$ [21]. Then, we say that $\mathcal{A}$ enumerates every solution of Π in $O(n^c)$ time for each. A *delay* of $\mathcal{A}$ is a computation time between two consecutive outputs. The algorithm $\mathcal{A}$ is *polynomial delay* if (1) the first solution is output in $O(n^c)$ time for some constant c and (2) the delay of $\mathcal{A}$ is bounded above by $O(n^c)$ [9]. Then, we say that, $\mathcal{A}$ enumerates every solution in $O(n^c)$ delay.

3 Family Tree of 2-Edge-Connected Induced Subgraphs

In this section, we define a tree structure among the set of 2-edge-connected induced subgraphs of an input graph. The vertices of the tree structure corresponds to the set of 2-edge-connected induced subgraphs, each edge corresponds to a parent-child relation between two 2-edge-connected induced subgraphs, and the root is the empty graph.

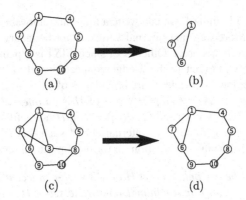

Fig. 2. (a) A 2-edge-connected induced subgraph H_1. (b) The parent $\mathcal{P}(H_1)$ of H_1. $\mathcal{P}(H_1)$ is obtained from H_1 by removing $\{4, 5, 8, 10, 9\}$, which is the smallest minimal removable set of H_1. (c) A 2-edge-connected induced subgraph H_2. (d) The parent $\mathcal{P}(H_2)$ of H_2. $\mathcal{P}(H_2)$ is obtained from H_2 by removing $\{3\}$, which is the smallest minimal removable set of H_2.

We define some notations. Let $G = (V(G), E(G))$ be a graph with a labeled-vertex set $V(G) = \{v_1, v_2, \ldots, v_n\}$ and an edge set $E(G)$. Let S_G be the set of the 2-edge-connected induced subgraphs of G, and let $C_G \subseteq S_G$ be the set of the induced cycles of G. We say that v_i is *smaller* than v_j, denoted by $v_i \prec v_j$, if $i < j$ holds. Let $H = (V(H), E(H))$ be a 2-edge-connected induced subgraph of G. Let $S_1 = \langle u_1, u_2, \ldots, u_k \rangle$, $(u_1 \prec u_k)$, and $S_2 = \langle w_1, w_2, \ldots, w_k \rangle$, $(w_1 \prec w_k)$, be two minimal removable sets of H. S_1 is smaller than S_2 if $u_1 \prec w_1$ holds. Note that, for two minimal removable sets S_1 and S_2, $S_1 \cap S_2 = \emptyset$ holds.

Let S be the smallest minimal removable set of H. Then, we define the *parent* of H, as follows.

$$\mathcal{P}(H) := \begin{cases} \emptyset & (H \in C_G) \\ H - S & (H \in S_G \backslash C_G), \end{cases}$$

where $\emptyset$ represents the empty graph, which is the graph with 0 vertex and 0 edge. We say that H is a *child* of the parent of $\mathcal{P}(H)$. Examples of parents are shown in Fig. 2. If $H \in S_G \backslash C_G$ holds, H has a removable set from Lemma 1. Hence, H always has its parent. Moreover, the parent is defined uniquely, since the smallest minimal removable set is unique in H. Note that $\mathcal{P}(H)$ is also a 2-edge-connected induced subgraph of G. For a 2-edge-connected induced subgraph in C_G, we define its parent as the empty graph $\emptyset$. By repeatedly finding parents from H, we obtain a sequence of 2-edge-connected induced subgraphs of G or the empty graph. We define the sequence $\mathcal{P}S(H) = \langle H_1, H_2, \ldots, H_\ell \rangle$, where $H_1 = H$ and $H_i = \mathcal{P}(H_{i-1})$ for $i = 2, 3, \ldots, \ell$, the *parent sequence* of H. An example of a parent sequence is shown in Fig. 3. This sequence ends up with the empty graph, as shown in the following lemma.

Lemma 5. *Let H be a 2-edge-connected induced subgraph of a graph G, and let $\mathcal{P}S(H) = \langle H_1, H_2, \ldots, H_\ell \rangle$ be the parent sequence of H. Then, H_ℓ is the empty graph $\emptyset$.*

Proof. Omitted.   □

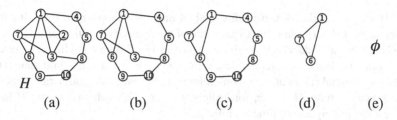

Fig. 3. An example of the parent sequence of a 2-edge-connected induced subgraph. (a) A 2-edge-connected induced subgraph H. (b) The parent $\mathcal{P}(H)$ of H. $\mathcal{P}(H)$ is obtained from H by removing $\{2\}$. (c) The parent $\mathcal{P}(\mathcal{P}(H))$ of $\mathcal{P}(H)$. $\mathcal{P}(\mathcal{P}(H))$ is obtained by removing $\{3\}$. (d) The parent $\mathcal{P}(\mathcal{P}(\mathcal{P}(H)))$ of $\mathcal{P}(\mathcal{P}(H))$. $\mathcal{P}(\mathcal{P}(\mathcal{P}(H)))$ is obtained by removing $\{4, 5, 8, 10, 9\}$. (e) Finally, the root, the empty graph, is obtained by removing $\{1, 6, 7\}$.

From Lemma 5, by merging the parent sequences of all 2-edge-connected induced subgraphs of G, we have the tree structure, called *family tree*, in which (1) the root is the empty graph, (2) the vertices except the root are 2-edge-connected induced subgraphs of G, and (3) each edge corresponds to a parent-child relation of two induced subgraphs of G. An example of the family tree is shown in Fig. 4.

4   Enumeration Algorithm

In this section, we present an enumeration algorithm for the 2-edge-connected induced subgraphs of an input graph. In the previous section, we defined the family tree rooted at the empty graph. Our algorithm enumerates 2-edge-connected induced subgraphs by traversing the tree. The children of the root in the tree are the induced cycles of an input graph. Our algorithm first enumerates the induced cycles of an input graph. Then, for each induced cycle, we traverse the subtree rooted at the cycle. To traverse the family tree, we have to design an enumeration algorithm for induced cycles of a graph and a child-enumeration algorithm for any 2-edge-connected induced subgraph. Fortunately, an efficient induced-cycle-enumeration algorithm is already known [20]. In our algorithm, we use the existing algorithm for enumerating the induced cycles of a graph. Now, in this section, we present a child-enumeration algorithm for 2-edge-connected induced subgraphs of a graph, below.

Let $G = (V(G), E(G))$ be a graph with a labeled-vertex set $V = \langle v_1, v_2, \ldots, v_n \rangle$ and an edge set E. Let $\mathcal{S}_G$ be the set of 2-edge-connected induced subgraphs of G, and let $\mathcal{C}_G \subseteq \mathcal{S}_G$ be the set of the induced cycles of G. To generate a child, we do the reverse operation for finding parents which is to attach a maximal internal ear to H. If the vertex set S of the attached path is the smallest minimal removable set in $H + S$, then $H + S$ is a child of H. Otherwise, $H + S$ is not a child. Let $\mathcal{I}(H, s, t)$ be the set of paths P such that P is a maximal internal ear in $H + P$ from s to t for $s, t \in N(H)$, where $N(H) := \bigcup_{u \in V(H)} N(u) \cap (V(G) \backslash V(H))$. For any different two $s, t \in N(H)$ and for any $P \in \mathcal{I}(H, s, t)$, $H + P$ is a candidate of a child, that is, $H + P$ may be a child. Therefore, if we generate all the paths in $\mathcal{I}(H, s, t)$ for every $s, t \in N(H)$, then all the children of H are enumerated by checking whether or not $H + P$ for each

$P \in \mathcal{I}(H, s, t)$ is a child. However, this method may take exponential time. It can be observed that $|\mathcal{I}(H, s, t)|$ can be an exponential of the number of vertices in $V(G)\backslash V(H)$ when $G - H$ has a "ladder" subgraph, as shown in Fig. 5 (one can choose to pass or not each rung in a ladder subgraph). Hence, there may be exponential child-candidates. If all the exponential candidates are non-children, the above child-enumeration method takes exponential time. However, fortunately, we can check whether or not H has at least one child in polynomial time, as follows.

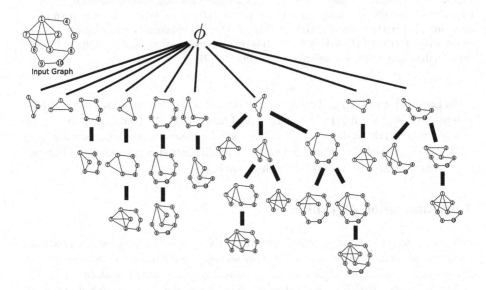

Fig. 4. An example of family tree of the input graph, drawn in the upper-left of the figure.

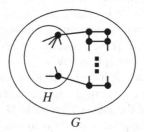

Fig. 5. A case that 2-edge-connected induced subgraph H of an input graph G has exponential child-candidates.

$\mathcal{M}(H)$ denotes the set of minimal removable sets of H. We can observe that, for any $P, P' \in \mathcal{I}(H, s, t)$, $\mathcal{M}(H + P) = \mathcal{M}(H + P')$ holds. Therefore, if any $P \in \mathcal{I}(H, s, t)$ is the smallest minimal removable set among $\mathcal{M}(H + P)$, then every $P' \in \mathcal{I}(H, s, t)\backslash\{P\}$

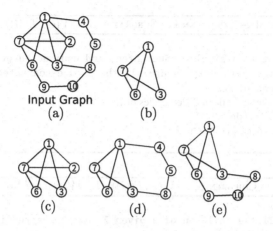

Fig. 6. Examples of children. (a) An input graph G. (b) A 2-edge-connected induced subgraph H of G. The smallest minimal removable set of H is $\{3\}$. (c) The induced subgraph obtained from H by inserting $\{2\}$. This is a child, since the smallest minimal removable set is $\{2\}$. (d) The induced subgraph obtained from H by inserting $\{4, 5, 6\}$. This subgraph is a child, since $\{4, 5, 6\}$ is the smallest minimal removable set. Note that, in this case, although $3 < 4$ holds, $\{3\}$ turns non-removable. (e) The induced subgraph obtained from H by inserting $\{8, 10, 9\}$. Here, the smallest minimal removable set is $\{1\}$. Hence, this is not a child. Note that, in this case, the set $\{1\}$ turns removable.

is also the smallest one among $\mathcal{M}(H + P')$. Therefore, if $H + P$ for $P \in \mathcal{I}(H, s, t)$ is a child of H, then $H + P'$ for every $P' \in \mathcal{I}(H, s, t) \backslash \{P\}$ is also a child of H. Hence, we focus on determining whether or not $H + P$ for a path $P \in \mathcal{I}(H, s, t)$, where $s, t \in N(H)$ and $s < t$, is a child of H. In the following case-analysis, we consider whether or not attaching $P \in \mathcal{I}(H, s, t)$ to H produces a child.

Case 1: $s < v$ for every $v \in V(H)$.

Obviously, for any path $P \in \mathcal{I}(H, s, t)$, $V(P)$ is the smallest minimal removable set in $H + P$. Therefore, $H + P$ is a child of H.

Case 2: Otherwise.

Let P be any path in $\mathcal{I}(H, s, t)$. If $V(P)$ is the smallest one among $\mathcal{M}(H + P)$, then $H + P$ is a child, and hence every $P' \in \mathcal{I}(H, s, t) \backslash \{P\}$ is also a child. Otherwise, $H + P$ is a non-child, and hence every $P' \in \mathcal{I}(H, s, t) \backslash \{P\}$ is also a non-child.

Examples of children are shown in Fig. 6. The figure includes three child-candidates of a 2-edge-connected induced subgraph.

Now, from the above case-analysis, we have the algorithm described in Algorithm 1 and Algorithm 2. In Algorithm 1, we are required to enumerate the induced cycles in an input graph G. This enumeration can be done using an existing algorithm [8], which enumerates all the induced cycles in C_G in $\tilde{O}(m + n |C_G|)$ time, where $\tilde{O}(f)$ is a shorthand for $O(f \cdot \text{polylog } n)$.

For each generated induced cycle, we traverse the subtree rooted at the cycle using Algorithm 2. In the line 10 in Algorithm 2, the algorithm constructs $\mathcal{M}(H + P)$. This

---

Algorithm 1: ENUM-2-EDGE-CONN-IND-SUBGRAPHS($G = (V(G), E(G))$)

---

1 **begin**
2   /* An input is a simple, unweighted, and undirected graph
      $G = (V(G), E(G))$. Outputs are all the 2-edge-connected induced
      subgraphs of G.   */
3   Let $C(G)$ be the set of the induced cycles of G.
4   **foreach** $C \in C(G)$ **do**
5   Call FIND-CHILDREN(G, C)

Algorithm 2: FIND-CHILDREN($G = (V(G), E(G))$, $H = (V(H), E(H))$)

---

1 **begin**
2   /* Find all the children of a given 2-edge-connected induced
      subgraph H.   */
3   Output H.
4   **foreach** $s, t \in N(H)$, $s < t$ **do**
5   **if** $s < v$ *for every* $v \in V(H)$ **then**
6   **foreach** $P \in I(H, s, t)$ **do**
7   FIND-CHILDREN($G, H + P$)
8   **else**
9   Let P be any path in $I(H, s, t)$.
10   Construct $M(H + P)$.
11   **if** $V(P)$ *is the smallest one among* $M(H + P)$ **then**
12   **foreach** $P \in I(H, s, t)$ **do**
13   FIND-CHILDREN($G, H + P$)

---

can be done, as follows. For each vertex u with degree 3 or more in $H + P$, we check whether or not $H + P - \{u\}$ is 2-edge-connected. If the answer is yes, $\{u\}$ is a member of $M(H + P)$. This check can be done in $O(m)$ time using a depth-first search. For each vertex u with degree 2 in $H + P$, we first find the maximal internal ear including u by traversing from u (This can be found in $O(n)$ time). Let P' be the found path. Then, we check whether or not $H + P - P'$ is 2-edge-connected. If the answer is yes, $V(P')$ is a member of $M(H + P)$. The above process can list all the members of $M(H + P)$ up and done in $O(nm) + O(nm)$. Note that the number of the internal ears in $H + P$ is bounded by $O(n)$. All the paths in $I(H, s, t)$ can be enumerated using an enumeration algorithm for induced paths [8], which enumerates all the paths in $I(H, s, t)$ in $\tilde{O}(m + n |I(H, s, t)|)$ time. Now, we estimate the running time of one recursive call of Algorithm 1. Suppose that H has k children. For each $O(n^2)$ pairs of s and t, we construct $M(H + P)$. This takes $O(n^3 m)$ time. Since H has k children, the enumeration algorithm for induced paths runs at most k times in lines 6 or 12. Let $\{s_1, t_1\}, \{s_2, t_2\}, \ldots, \{s_x, t_x\}$ be pairs of vertices in $N(H)$ such that, for $P \in I(H, s_i, t_i)$, $H + P$ is a child of H. Let $k_i = |I(H, s_i, t_i)|$. Note that there exists k_i children generated by attaching $P \in I(H, s_i, t_i)$. Then, $k = \sum_{1 \leq i \leq x} k_i$

holds. Now, the total running time for enumerating paths in $I(H, s_i, t_i)$ is bounded by $\sum_{1 \le i \le x} \tilde{O}(m + nk_i)$, which is bounded by $k \cdot \tilde{O}(m + n)$. Hence, each recursive call of Algorithm 2 takes at most $O(n^3 m) + k \cdot \tilde{O}(m + n)$ time. Therefore, we have the following theorem.

Theorem 3. *Let $G = (V(G), E(G))$ be a graph with n vertices and m edges. Let S_G be the set of the 2-edge-connected induced subgraphs of G. One can enumerate all the 2-edge-connected induced subgraphs of G in $O(n^3 m |S_G|)$ time.*

Proof. All the children of the root of the family tree, namely the induced cycles of G, can be enumerated in in $\tilde{O}(m + n |C_G|)$ time [8]. The recursive call for any 2-edge-connected induced subgraph H takes $O(n^3 m) + k \cdot \tilde{O}(m + n)$ if H has k children. Hence, the total running time of our algorithm is bounded by $O(n^3 m |S_G|)$.   □

Using alternative output technique [15], we have a polynomial delay enumeration algorithm for 2-edge-connected induced subgraphs. First, we enumerate induced cycles in polynomial delay using the algorithm [20]. Then, for each induced cycle, we traverse the subtree rooted at the cycle. In that traversal, we apply the alternative output technique. In the technique, 2-edge-connected induced subgraphs with even-depth in the tree are output before their children and 2-edge-connected induced subgraphs with odd-depth in the tree are output after their children. In the above traversal, we have an output per at most 3 edge traversals in each subtree. Hence, we have the following corollary.

Corollary 1. *Let $G = (V(G), E(G))$ be a graph with n vertices and m edges. One can enumerate every 2-edge-connected induced subgraph in polynomial delay.*

References

1. Akiba, T., Iwata, Y., Yoshida, Y.: Linear-time enumeration of maximal k-edge-connected subgraphs in large networks by random contraction. In: Proceedings of the 22nd ACM International Conference on Information and Knowledge Management, CIKM 2013, pp. 909–918 (2013)
2. Avis, D., Fukuda, K.: Reverse search for enumeration. Discret. Appl. Math. **65**(1–3), 21–46 (1996)
3. Birmelé, E., et al.: Optimal listing of cycles and st-paths in undirected graphs. In: Proceedings of the 24th Annual ACM-SIAM Symposium on Discrete Algorithms, SODA 2012, pp. 1884–1896 (January 2012)
4. Boros, E., Borys, K., Elbassioni, K.M., Gurvich, V., Makino, K., Rudolf, G.: Generating minimal k-vertex connected spanning subgraphs. In: Proceedings of the 13th Annual International Computing and Combinatorics Conference, COCOON 2007, pp. 222–231 (2007)
5. Conte, A., Kanté, M.M., Otachi, Y., Uno, T., Wasa, K.: Efficient enumeration of maximal k-degenerate subgraphs in a chordal graph. In: Proceedings of the 23rd Annual International Computing and Combinatorics Conference, COCOON 2017, pp. 150–161 (2017)
6. Conte, A., Uno, T.: New polynomial delay bounds for maximal subgraph enumeration by proximity search. In: Proceedings of the 51st Annual ACM SIGACT Symposium on Theory of Computing, STOC 2019, pp. 1179–1190 (2019)

7. Conte, A., De Virgilio, R., Maccioni, A., Patrignani, M., Torlone, R.: Finding all maximal cliques in very large social networks. In: Proceedings of the 19th International Conference on Extending Database Technology, pp. 173–184 (2016)

8. Ferreira, R., Grossi, R., Rizzi, R., Sacomoto, G., Sagot, M.-F.: Amortized $\tilde{O}(—V—)$-delay algorithm for listing chordless cycles in undirected graphs. In: Schulz, A.S., Wagner, D. (eds.) ESA 2014. LNCS, vol. 8737, pp. 418–429. Springer, Heidelberg (2014). https://doi.org/10.1007/978-3-662-44777-2_35

9. Johnson, D.S., Papadimitriou, C.H., Yannakakis, M.: On generating all maximal independent sets. Inf. Process. Lett. **27**(3), 119–123 (1988)

10. Khachiyan, L., Boros, E., Borys, K., Elbassioni, K., Gurvich, V., Makino, K.: Enumerating spanning and connected subsets in graphs and matroids. In: Azar, Y., Erlebach, T. (eds.) ESA 2006. LNCS, vol. 4168, pp. 444–455. Springer, Heidelberg (2006). https://doi.org/10.1007/11841036_41

11. Kurita, K., Wasa, K., Uno, T., Arimura, H.: Efficient enumeration of induced matchings in a graph without cycles with length four. IEICE Trans. Fundam. Electron. Commun. Comput. Sci. **101–A**(9), 1383–1391 (2018)

12. Makino, K., Uno, T.: New algorithms for enumerating all maximal cliques. In: Hagerup, T., Katajainen, J. (eds.) SWAT 2004. LNCS, vol. 3111, pp. 260–272. Springer, Heidelberg (2004). https://doi.org/10.1007/978-3-540-27810-8_23

13. Marino, A.: Analysis and Enumeration. Atlantis Press (2015)

14. Maxwell, S., Chance, M.R., Koyutürk, M.: Efficiently enumerating all connected induced subgraphs of a large molecular network. In: Dediu, A.-H., Martín-Vide, C., Truthe, B. (eds.) AlCoB 2014. LNCS, vol. 8542, pp. 171–182. Springer, Cham (2014). https://doi.org/10.1007/978-3-319-07953-0_14

15. Nakano, S., Uno, T.: Generating colored trees. In: Kratsch, D. (ed.) WG 2005. LNCS, vol. 3787, pp. 249–260. Springer, Heidelberg (2005). https://doi.org/10.1007/11604686_22

16. Read, R.C., Tarjan, R.E.: Bounds on backtrack algorithms for listing cycles, paths, and spanning trees. Networks **5**(3), 237–252 (1975)

17. Shioura, A., Tamura, A., Uno, T.: An optimal algorithm for scanning all spanning trees of undirected graphs. SIAM J. Comput. **26**(3), 678–692 (1997)

18. Uno, T.: An efficient algorithm for solving pseudo clique enumeration problem. Algorithmica **56**(1), 3–16 (2010)

19. Uno, T.: Constant time enumeration by amortization. In: Dehne, F., Sack, J.-R., Stege, U. (eds.) WADS 2015. LNCS, vol. 9214, pp. 593–605. Springer, Cham (2015). https://doi.org/10.1007/978-3-319-21840-3_49

20. Uno, T., Satoh, H.: An efficient algorithm for enumerating chordless cycles and chordless paths. In: Džeroski, S., Panov, P., Kocev, D., Todorovski, L. (eds.) DS 2014. LNCS (LNAI), vol. 8777, pp. 313–324. Springer, Cham (2014). https://doi.org/10.1007/978-3-319-11812-3_27

21. Valiant, L.G.: The complexity of computing the permanent. Theoret. Comput. Sci. **8**, 189–201 (1979)

22. Wasa, K., Kaneta, Y., Uno, T., Arimura, H.: Constant time enumeration of subtrees with exactly k nodes in a tree. IEICE Trans. Inf. Syst. **97–D**(3), 421–430 (2014)

23. West, D.B.: Introduction to Graph Theory, 2nd edn. Prentice Hall (September 2000)

24. Yamanaka, K., Matsui, Y., Nakano, S.-I.: Enumerating highly-edge-connected spanning subgraphs. IEICE Trans. Fundame. Electron. Commun. Comput. Sci. **102–A**(9), 1002–1006 (2019)

An Optimal Algorithm for BISECTION for Bounded-Treewidth Graph

Tesshu Hanaka[1] [iD], Yasuaki Kobayashi[2(✉)] [iD], and Taiga Sone[2]

[1] Chuo University, Kasuga, Bunkyo-ku, Tokyo 112-8551, Japan
hanaka.91t@g.chuo-u.ac.jp
[2] Kyoto University, Yoshida-honmachi, Sakyo-ku, Kyoto 606-8501, Japan
{kobayashi,taiga_sone}@iip.ist.i.kyoto-u.ac.jp

Abstract. The maximum/minimum bisection problems are, given an edge-weighted graph, to find a bipartition of the vertex set into two sets whose sizes differ by at most one, such that the total weight of edges between the two sets is maximized/minimized. Although these two problems are known to be NP-hard, there is an efficient algorithm for bounded-treewidth graphs. In particular, Jansen et al. (SIAM J. Comput. 2005) gave an $O(2^t n^3)$-time algorithm when given a tree decomposition of width t of the input graph, where n is the number of vertices of the input graph. Eiben et al. (ESA 2019) improved the running time to $O(8^t t^5 n^2 \log n)$. Moreover, they showed that there is no $O(n^{2-\varepsilon})$-time algorithm for trees under some reasonable complexity assumption.

In this paper, we show an $O(2^t (tn)^2)$-time algorithm for both problems, which is asymptotically tight to their conditional lower bound. We also show that the exponential dependency of the treewidth is asymptotically optimal under the Strong Exponential Time Hypothesis. Moreover, we discuss the (in)tractability of both problems with respect to special graph classes.

Keywords: Bisection · Fixed-parameter tractable · Treewidth

1 Introduction

Let $G = (V, E)$ be a graph and let $w : E \to \mathbb{R}$ be an edge-weight function. For disjoint subsets X, Y of V, we denote by $w(X, Y)$ the total weight of edges between X and Y. A *bisection* of G is a bipartition of V into two sets A and B such that $-1 \le |A| - |B| \le 1$. The *size* of a bisection (A, B) is defined as the number of edges between A and B. We also consider bisections of edge-weighted graphs. In this case, the size of a bisection (A, B) is defined as $w(A, B)$.

In this paper, we consider the following two problems: MIN BISECTION and MAX BISECTION.

This work is partially supported by JSPS KAKENHI Grant Numbers JP19K21537, JP17H01788 and JST CREST JPMJCR1401.

M. Li (Ed.): FAW 2020, LNCS 12340, pp. 25–36, 2020.
https://doi.org/10.1007/978-3-030-59901-0_3

MIN BISECTION ────────────────────────────────────

Input: a graph $G = (V, E)$ and $w : E \to \mathbb{R}_+$.
Goal: Find a minimum size bisection (A, B) of G.

──

MAX BISECTION ────────────────────────────────────

Input: a graph $G = (V, E)$ and $w : E \to \mathbb{R}_+$.
Goal: Find a maximum size bisection (A, B) of G.

──

If the edge weight is allowed to be arbitrary, the problems are equivalent to each other. These problems are well-known variants of MINCUT and MAXCUT. If every edge has non-negative weight, MINCUT, which is the problem of minimizing $w(A, B)$ over all bipartition (A, B) of V, can be solved in polynomial time. For MAXCUT, which is the maximization version of MINCUT, the problem is NP-hard in general [13] and trivially solvable in polynomial time for bipartite graphs. Orlova and Dorfman [17] and Hadlock [10] proved that MAXCUT can be solved in polynomial time for planar graphs with non-negative edge weights, and Shih et al. [20] finally gave a polynomial-time algorithm for planar graphs with arbitrary edge weights. However, the complexity of bisection problems is quite different from MINCUT and MAXCUT. MAX BISECTION is known to be NP-hard even for planar graphs [12] and unit disk graphs [5]. For MIN BISECTION, it is known to be NP-hard [8] even for d-regular graphs for fixed $d \geq 3$ [3] and unit disk graphs [6]. It is worth noting that MIN BISECTION for planar graphs is still open.

On bounded-treewidth graphs, MIN BISECTION and MAX BISECTION are solvable in polynomial time. More precisely, given an input graph G of n vertices and a tree decomposition of G of width t, Jansen et al. [12] proved that MAX BISECTION can be solved in time $O(2^t n^3)$. This algorithm also works on graphs with arbitrary edge-weights, which means MIN BISECTION can be solved within the same running time. Very recently, Eiben et al. [7] improved the polynomial factor of n with a sacrifice of the exponential factor in t in the running time. The running time of their algorithm is $O(8^t t^5 n^2 \log n)$. They also discussed a conditional lower bound on the running time: They showed that for any $\varepsilon > 0$, MIN BISECTION cannot be solved in $O(n^{2-\varepsilon})$ for n-vertex trees unless (min, +)-CONVOLUTION (defined in Sect. 3.3) has an $O(n^{2-\delta})$-time algorithm for some $\delta > 0$. Since trees are precisely connected graphs of treewidth at most one, this lower bound also holds for bounded-treewidth graphs. Therefore, there is still a gap between the upper and (conditional) lower bound on the running time for bounded-treewidth graphs.

In this paper, we give an "optimal" algorithm for MIN BISECTION and MAX BISECTION on bounded-treewidth graphs. The running time of our algorithm is $O(2^t(tn)^2)$, provided that a width-t tree decomposition of the input graph is given as input. The polynomial factor in n matches the conditional lower bound given by Eiben el al. [7]. We also show that MAX BISECTION cannot be solved in time $(2 - \varepsilon)^t n^{O(1)}$ for any $\varepsilon > 0$ unless the Strong Exponential Time Hypothesis (SETH) [11] fails. These facts imply that the exponential dependency with respect to t and the polynomial dependency with respect to n in our running time are asymptotically optimal under these well-studied complexity-theoretic assumptions.

We also investigate the complexity of MIN BISECTION and MAX BISECTION on special graph classes. From the hardness of MAXCUT, we immediately have several complexity results for MAX BISECTION and MIN BISECTION. The most notable case is that both problems are NP-hard even on unweighted bipartite graphs, on which MAXCUT can be trivially solved in polynomial time. Apart from these complexity results, we show that MAX BISECTION can be solved in linear time on line graphs.

2 Preliminaries

Let $G = (V, E)$ be a graph, which is simple and undirected. Throughout the paper, we use n to denote the number of vertices of an input graph. For a vertex $v \in V$, we denote by $N(v)$ the set of neighbors of v in G. For two disjoint subsets $X, Y \subseteq V$, we denote by $E(X, Y)$ the set of edges having one end in X and the other end in Y. We write $w(X, Y)$ to denote the total weight of edges in $E(X, Y)$ (i.e., $w(X, Y) = \sum_{e \in E(X,Y)} w(e)$). A bipartition (A, B) of V is called a *cut* of G. The *size* of a cut is the number of edges between A and B, that is, $|E(A, B)|$. For edge-weighted graphs, the size is measured by the total weight of edges between A and B. A cut is called a *bisection* if $-1 \le |A| - |B| \le 1$.

In the next section, we work on dynamic programming based on *tree decompositions*. A *tree decomposition* of G is a pair of a rooted tree T with vertex set I and a collection $\{X_i : i \in I\}$ of subsets of V such that

- $\bigcup_{i \in I} X_i = V$;
- for each $\{u, v\} \in E$, there is an $i \in I$ with $\{u, v\} \subseteq X_i$;
- for each $v \in V$, the subgraph of T induced by $\{i \in I : v \in X_i\}$ is connected.

We refer to the vertices of T as *nodes* to distinguish them from vertices of G. The *width* of T is defined as $\max_{i \in I} |X_i| - 1$. The *treewidth* of G is the minimum integer k such that G has a tree decomposition of width k.

To facilitate dynamic programming on tree decompositions, several types of "special" tree decompositions are known. Jansen et al. [12] used the well-known *nice tree decomposition* for solving MAX BISECTION. Eiben et al. [7] improved the dependency on n by means of "shallow tree decompositions" due to Bodlaender and Hagerup [1]. In this paper, we rather use nice tree decompositions as well as Jansen et al. [12], and the algorithm itself is, in fact, identical with theirs.

We say that a tree decomposition T is *nice* if for every non-leaf node i of T, either

- **Introduce node** i has an exactly one child $j \in I$ such that $X_i = X_j \cup \{v\}$ for some $v \in V \setminus X_j$,
- **Forget node** i has an exactly one child $j \in I$ such that $X_j = X_i \cup \{v\}$ for some $v \in V \setminus X_i$, or
- **Join node** i has exactly two children $j, k \in I$ such that $X_i = X_j = X_k$.

Lemma 1 (Lemma 13.1.3 in [14]). *Given a tree decomposition of G of width t, there is an algorithm that converts it into a nice tree decomposition of width at most t in time $O(t^2 n)$. Moreover, the constructed nice tree decomposition has at most $4n$ nodes.*

3   Bounded-Treewidth Graphs

Let $G = (V, E)$ be an edge-weighted graph with weight function $w : E \to \mathbb{R}$. Note that we do not restrict the weight function to take non-negative values. Given this, MIN BISECTION is essentially equivalent to MAX BISECTION. Therefore, in this section, we will only consider the maximization counterpart.

3.1   An $O(2^t n^3)$-Time Algorithm

In this subsection, we quickly review the algorithm of Jansen et al. [12]. Let T be a nice tree decomposition of width at most t. For each node $i \in I$, we use V_i to denote the set of vertices of G that is contained in X_i or X_j for some descendant $j \in I$ of i.

Let $i \in I$ be a node of T. For each $S \subseteq X_i$ and $0 \le d \le |V_i|$, we compute the value $\mathrm{bs}(i, S, d)$ which is the maximum size of a bisection (A_i, B_i) of $G[V_i]$ such that $A_i \cap X_i = S$ and $|A_i| = d$.

Leaf Node: Let $i \in I$ be a leaf of T. For each $S \subseteq X_i$, $\mathrm{bs}(i, S, d) = w(S, X_i \setminus S)$ if $d = |S|$. Otherwise we set $\mathrm{bs}(i, S, d) = -\infty$.

Introduce Node: Let $i \in I$ be an introduce node of T and let $v \in X_i \setminus X_j$ be the vertex introduced at i, where $j \in I$ is the unique child of i. Since the neighborhood of v in $G[V_i]$ is entirely contained in X_i, we can compute $\mathrm{bs}(i, S, d)$ as

$$\mathrm{bs}(i, S, d) = \begin{cases} \mathrm{bs}(j, S \setminus \{v\}, d - 1) + w(\{v\}, X_i \setminus S) & \text{if } v \in S \\ \mathrm{bs}(j, S, d) + w(\{v\}, S) & \text{otherwise,} \end{cases}$$

for each $S \subseteq X_i$ and $0 \le d \le |V_i|$.

Forget Node: Let $i \in I$ be a forget node of T and let $v \in X_j \setminus X_i$ be the vertex forgotten at i, where $j \in I$ is the unique child of i. As $G[V_i] = G[V_j]$, we can compute $\mathrm{bs}(i, S, d)$ as

$$\mathrm{bs}(i, S, d) = \max(\mathrm{bs}(j, S, d), \mathrm{bs}(j, S \cup \{v\}, d))$$

for each $S \subseteq X_i$ and $0 \le d \le |V_i|$.

Join Node: Let $i \in I$ be a join node of T with children $j, k \in I$. By the definition of nice tree decompositions, we have $X_i = X_j = X_k$. For $S \subseteq X_i$ and $0 \le d \le |V_i|$,

$$\text{bs}(i, S, d) = \max_{|S| \le d' \le d} (\text{bs}(j, S, d') + \text{bs}(k, S, d - d' + |S|) - w(S, X_i \setminus S)). \quad (1)$$

Note that the edges between S and $X_i \setminus S$ contribute to both $\text{bs}(j, S, d')$ and $\text{bs}(k, S, d - d' + |S|)$. Thus, we subtract $w(S, X_i \setminus S)$ in the recurrence (1).

Running Time: For each Leaf/Introduce/Forget node i, we can compute $\text{bs}(i, S, d)$ in total time $O(2^t t |V_i|)$ for all $S \subseteq X_i$ and $0 \le d \le |V_i|$. For join node i, the recurrence (1) can be evaluated in $O(|V_i|)$ time for each $S \subseteq X_i$ and $0 \le d \le |V_i|$, provided that $\text{bs}(j, S, *)$, $\text{bs}(k, S, *)$, and $w(S, X_i \setminus S)$ are stored in the table. Therefore, the total running time for a join node i is $O(2^t |V_i|^2)$. Since $|V_i| = O(n)$ and T has $O(n)$ nodes, the total running time of the entire algorithm is $O(2^t n^3)$.

Theorem 1 ([12]). *Given a tree decomposition of G of width t, MAX BISECTION can be solved in $O(2^t n^3)$ time.*

3.2 A Refined Analysis for Join Nodes

For a refined running time analysis, we reconsider the recurrence (1) for join nodes. This can be rewritten as

$$\begin{aligned}
\text{bs}(i, S, d) &= \max_{|S| \le d' \le d} (\text{bs}(j, S, d') + \text{bs}(k, S, d - d' + |S|) - w(S, X_i \setminus S)) \\
&= \max_{\substack{d', d'' \\ d' + d'' = d + |S|}} (\text{bs}(j, S, d') + \text{bs}(k, S, d'') - w(S, X_i \setminus S)).
\end{aligned}$$

Since d' and d'' respectively run over $0 \le d' \le |V_j|$ and $0 \le d'' \le |V_k|$, we can compute $\text{bs}(i, S, d)$ in total time $O(2^t |V_j| \cdot |V_k|)$ for all S and d.

For each node $i \in I$, we let $n_i = \sum_{j \preceq_T i} |X_j|$, where the summation is taken over all descendants j of i and i itself. Clearly, $n_i \ge |V_i|$ and hence the total running time of join nodes is upper bounded by

$$\sum_{i: \text{ join node}} O(2^t n_j n_k) = O\left(2^t \cdot \sum_{i: \text{ join node}} n_j n_k\right).$$

Note that we abuse the notations n_j and n_k for different join nodes i, and the children nodes j and k are defined accordingly. We claim that $\sum_{i: \text{ join node}} n_j n_k$ is $O((tn)^2)$. To see this, let us consider the term $n_j n_k$ for a join node i. For each node l of T, we label all the vertices contained in X_l by distinct labels $v_1^l, v_2^l, \ldots v_{|X_l|}^l$. Note that some single vertex can receive two or more labels since a vertex can be contained in more than one node in the tree decomposition.

From now on, we regard such a vertex as distinct labeled vertices and hence n_i corresponds to the number of labeled vertices that appear in the node i or some descendant node of i. Now, the term $n_j n_k$ can be seen as the number of pairs of labeled vertices (l, r) such that l is a labeled vertex contained in the subtree rooted at the left child j and r is a labeled vertex contained in the subtree rooted at the right child k. A crucial observation is that for any pair of labeled vertices, there is at most one join node i counted it in the term $n_j n_k$. This implies that $\sum_{i: \text{ join node}} n_j n_k$ is at most the number of pairs of distinct labeled vertices. Since each node of T contains at most $t + 1$ vertices and T contains $O(n)$ nodes, it is $O((tn)^2)$. Therefore, the total running time of the algorithm is $O(2^t(tn)^2)$.

Theorem 2. *Given a tree decomposition of G of width t, MAX BISECTION can be solved in time $O(2^t(tn)^2)$.*

3.3   Optimality of Our Algorithm

Eiben et al. [7] proved that if the following $(\min, +)$-CONVOLUTION does not admit $O(n^{2-\delta})$-time algorithm for some $\delta > 0$, there is no $O(n^{2-\varepsilon})$-time algorithm for MIN BISECTION on trees for any $\varepsilon > 0$.

$(\min, +)$-CONVOLUTION
Input: Two sequences of numbers $(a_i)_{1 \leq i \leq n}$, $(b_i)_{1 \leq i \leq n}$.
Goal: Compute $c_i = \min_{1 \leq j \leq i}(a_j + b_{i-j+1})$ for all $1 \leq i \leq n$.

This conditional lower bound matches the dependency on n in the running time of our algorithm.

In terms of the dependency on treewidth, we can prove that under the Strong Exponential Time Hypothesis (SETH) [11], and hence the exponential dependency on t asymptotically optimal.

Theorem 3 ([16]). *Unless SETH fails, there is no algorithm for the unweighted maximum cut problem that runs in time $2^{t-\varepsilon}n^{O(1)}$ for any $\varepsilon > 0$ even if a width-t tree decomposition of the input graph is given as input for some t.*

The known reduction (implicitly appeared in [3]) from the unweighted MAX-CUT to MAX BISECTION works well for our purpose. Specifically, let G be an unweighted graph and let n be the number of vertices of G. We add n isolated vertices to G and the obtained graph is denoted by G'. It is easy to see that G has a cut of size at least k if and only if G' has a bisection of size at least k. Moreover, $\text{tw}(G') = \text{tw}(G)$. Therefore, we have the following lower bound.

Theorem 4. *Unless SETH fails, there is no algorithm for MAX BISECTION that runs in time $2^{t-\varepsilon}n^{O(1)}$ for any $\varepsilon > 0$ even if a width-t tree decomposition of the input graph is given as input for some t.*

4   Hardness on Graph Classes

In this section, we discuss some complexity results for MIN BISECTION and MAX BISECTION. In Sect. 3.3, we have seen that there is a quite simple reduction from MAXCUT to MAX BISECTION. We formally describe some immediate consequences of this reduction as follows. Let $\mathcal{C}$ be a graph class such that

- MAXCUT is NP-hard even if the input graph is restricted to be in C and
- for every $G \in C$, a graph G' obtained from G by adding arbitrary number of isolated vertices is also contained in C.

The reduction shows that MAX BISECTION is NP-hard for every graph class C that satisfies the above conditions.

Theorem 5. MAX BISECTION *is NP-hard even for split graphs, comparability graphs, AT-free graphs, and claw-free graphs.*

It is known that MAXCUT is NP-hard even for split graphs [2] and comparability graphs [18], and co-bipartite graphs [2] that is a subclass of AT-free graphs and claw-free graphs. If C is the class of co-bipartite graphs, the second condition does not hold in general. However, we can prove the hardness of MAX BISECTION on co-bipartite graphs, which will be discussed in the last part of this section.

Suppose the input graph G has $2n$ vertices. Let $\overline{G}$ is the complement of G. It is easy to see that G has a bisection of size at least k if and only if $\overline{G}$ has a bisection of size at most $n^2 - k$. This immediately gives the following theorem from Theorem 5.

Theorem 6. MIN BISECTION *is NP-hard even for split graphs and co-comparability graphs.*

For bipartite graphs, MAX CUT is solvable in polynomial time. However, we show that MIN BISECTION and MAX BISECTION are NP-hard even on bipartite graphs.

Theorem 7. MIN BISECTION *is NP-hard even for bipartite graphs.*

Proof. We prove the theorem by performing a polynomial-time reduction from MIN BISECTION on 4-regular graphs, which is known to be NP-hard [3].

Let $G = (V, E)$ be a 4-regular graph. We can assume that G has $2n$ vertices since the reduction given by [3] works on graphs having even number of vertices. For each edge $e = \{u, w\} \in E$, we split e by introducing a new vertex v_e and replacing e with two edges $\{u, v_e\}$ and $\{v_e, w\}$. Then, for each $v \in V$, we add n^3 pendant vertices and make adjacent them to v. We denote by V_E the set of vertices newly added for edges, by V^p the set of pendant vertices, and by $G' = (V \cup V_E \cup V^p, E')$ the graph obtained from G as above (see Fig. 1). As G is 4-regular, we have $|V_E| = |E| = 4n$ and $|V \cup V_E \cup V^p| = 2n + 4n + 2n^4 = 2n^4 + 6n$. Moreover, G' is bipartite. In the following, we show that G has a bisection of size at most k if and only if so does G'.

Suppose G has a bisection (V_1, V_2) of size at most k. Since $|V| = 2n$, it holds that $|V_1| = |V_2|$. For $i = 1, 2$, we set $V_i' = V_i \cup \{v_e : e \subseteq V_i\} \cup V_i^p$, where V_i^p is the set of pendant vertices such that its unique neighbor is contained in V_i. Note that there are no edges between V_1' and V_2' in G' and $|V_1'| = |V_2'|$ so far. Observe that for every $e \in E(V_1, V_2)$, exactly one of the incidental edges $\{u, v_e\}$ and $\{v_e, w\}$ of the corresponding vertex v_e contributes to its size no matter whether

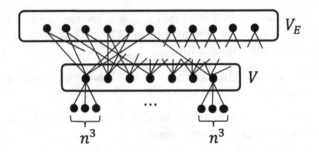

Fig. 1. The constructed graph in the proof of Theorem 7.

v_e is included in either V_1' or V_2'. Therefore, we can appropriately distribute the remaining vertices $\{v_e : e \in E(V_1, V_2)\}$ to obtain a bisection of size at most k.

Suppose that G' has a bisection (V_1', V_2') of size at most k. Let $V_1 = V_1' \cap V$ and $V_2 = V_2'$. We claim that $|V_1| = |V_2|$. Suppose for contradiction that $|V_1| > |V_2|$. As $|V_1 \cup V_2| = 2n$, we have $|V_1| \geq n+1$. Since (V_1', V_2') is a bisection of G', it holds that $|V_1'| = n^4 + 3n$. Thus, there are at least $n^3 - 2n + 1$ pendant vertices in V_2' whose neighbor is contained in V_1'. For $n \geq 5$, it holds that $n^3 - 2n + 1 > 4n^2 > k$, contradicting to the assumption that (V_1', V_2') is a bisection of size at most k. Moreover, for every edge $e = \{u, w\} \in E(V_1, V_2)$ in G, at least one of $\{u, v_e\}$ or $\{v_e, w\}$ contributes to the cut (V_1', V_2') in G'. Therefore, we conclude that the size of the cut (V_1, V_2) in G is at most k.   □

Interestingly, the same construction works well for proving the hardness of Max Bisection for bipartite graphs.

Theorem 8. Max Bisection *is NP-hard even for bipartite graphs.*

Proof. The proof of this theorem is similar to that of Theorem 7. Let $G = (V, E)$ be a 4-regular graph and let $G' = (V \cup V_E \cup V^p, E')$ be the bipartite graph described in the proof of Theorem 7. In the following, we prove that G has a bisection of size at most k if and only if G' has a bisection of size *at least* $2n^4 + 8n - k$.

Suppose first that G has a bisection (V_1, V_2) of size k. Then, we set V_i' for $i \in \{1, 2\}$ as:

- $V_i \subseteq V_i'$;
- If $v \in V_i$, all the pendant vertices w with $N(w) = \{v\}$ are contained in V_{3-i}';
- For each $e \in E$ with $e \subseteq V_i$, v_e is contained in V_{3-i}'.

For each remaining $e \in E(V_1, V_2)$, we add v_e to arbitrary side V_i' so that (V_1', V_2') becomes a bisection of G'. This can be done since G is 4-regular, which means $G[V_i]$ contains exactly $2n - k$ edges for each $i = \{1, 2\}$. Let us note that $E'(V_1', V_2')$ has $2n^4 + 8n - 2k$ edges and for $v_e \in E$ with $e \in E(V_1, V_2)$, exactly one of the incident edges of v_e contributes to the size of the bisection no matter which V_i' includes v_e. This implies that the size of bisection (V_1', V_2') is $2n^4 + 8n - k$.

For the converse, suppose that G' has a bisection (V_1', V_2') of size at least $2n^4 + 8n - k$. For each $i = 1, 2$, we let $V_i = V_i' \cap V$. Then, we claim that (V_1, V_2) is a bisection of G. To see this, we assume for contradiction that $|V_1| > |V_2|$. Clearly, V_1 contains at least $n + 1$ vertices. As $|V_1'| = |V_2'|$ and G' has $2n + 2n \cdot n^3 + 4n = 2n^4 + 6n$ vertices, we have $|V_2'| = n^4 + 3n$. Since V_1 has at least $n + 1$ vertices, at least $(n+1)n^3 - |V_2'| = n^3 - 3n$ pendant vertices adjacent to some vertex in V_1 are included in V_1. Therefore, at most $|E'| - (n^3 - 3n) = 2|E| + 2n \cdot n^3 - (n^3 - 3n) = 2n^4 - n^3 + 12n$ edges can belong to $E'(V_1', V_2')$. For $n \geq 3$, we have $2n^4 - n^3 + 12n < 2n^4 + 8n - 4n < 2n^4 + 8n - k$, which contradicts to the fact that the size of (V_1', V_2') is at least $2n^4 + 8n - k$. Note that $k \leq 4n$.

Now, we show that the bisection (V_1, V_2) of G has size at most k. Since there are $2n^4$ pendant edges in G', at least $8n - k$ edges of $G'[V \cup V_E]$ belong to $E'(V_1', V_2')$. Note that as V and V_E are respectively independent sets in G', such edges are in $E'(V, V_E)$. Moreover, there are $8n$ edges in $E'(V, V_E)$. If there are at least $k + 1$ vertices v_e in V_E having neighbors both in V_1 and in V_2, the size of $E'(V_1' \cap (V \cup V_E), V_2' \cap (V \cup V_E))$ is at most $8n - k - 1$ since exactly one of the incidental edges of v_e does not contribute to the cut. Thus, the number of such vertices is at most k. Since each $v_e \in V_E$ having neighbors both in V_1 and in V_2 corresponds to a cut edge of (V_1, V_2) in G, the size of the bisection (V_1, V_2) of G is at most k.   □

Since both MIN BISECTION and MAX BISECTION are NP-hard on bipartite graphs, by the same argument with Theorem 6, we have the following corollary.

Corollary 1. MIN BISECTION *and* MAX BISECTION *are NP-hard even for co-bipartite graphs.*

5   Line Graphs

Guruswami [9] showed that MAXCUT can be solved in linear time for unweighted line graphs. The idea of the algorithm is to find a cut satisfying a certain condition using an Eulerian tour of the underlying graph of the input line graph. In this section, we show that his approach works well for MAX BISECTION.

Let $G = (V, E)$ be a graph. The *line graph* of G, denoted by $L(G) = (V_L, E_L)$, is an undirected graph with $V_L = E$ such that two vertices $e, f \in V_L$ are adjacent to each other if and only if e and f share their end vertex in G. We call G an *underlying graph* of $L(G)$. Note that from a line graph, its underlying graph is not uniquely determined. However, it is sufficient to take an arbitrary one of them to discuss our result. Guruswami gave the following sufficient condition for MAXCUT and showed that every line graph has a cut satisfying this condition.

Lemma 2 ([9]). *Let* $G = (V, E)$ *be a (not necessarily line) graph and let* $C_1, C_2, \ldots C_k$ *be edge disjoint cliques with* $\bigcup_{1 \leq i \leq k} C_i = E$. *If there is a cut* (A, B) *of* G *such that* $-1 \leq |A \cap C_i| - |B \cap C_i| \leq 1$ *for every* $1 \leq i \leq k$, *then* (A, B) *is a maximum cut of* G.

Since the maximum size of a bisection is at most the maximum size of a cut, we immediately conclude that every bisection satisfying the condition in Lemma 2 is a maximum bisection. The construction of a bipartition (A, B) of V in [9] is as follows.

Let $L(G) = (V_L, E_L)$ be a line graph whose underlying graph is G. We make G an even-degree graph by putting a vertex r and make adjacent r to each vertex of odd degree. Let G' be the even-degree graph obtained as above. Suppose first that G' is connected. Fix an Eulerian tour starting from r and alternately assign labels a and b to each edge along with the Eulerian tour. Let A and B be the set of edges having label a and b, respectively. Observe that the bipartition $(A \cap V_L, B \cap V_L)$ of V_L satisfies the sufficient condition in Lemma 2. To see this, consider a vertex v of G. Since the set of edges C_v adjacent to v forms a clique in $L(G)$. Moreover, it is known that, in line graphs, the edges of cliques $\{C_v : v \in G\}$ partitions the whole edge set E_L. Every vertex v of G' except for r has an equal number of incidental edges with label a and those with label b in G', which implies that every clique C_v satisfies $-1 \leq |C_v \cap (A \cap V_L)| - |C_v \cap (B \cap V_L)| \leq 1$.

Now, we show that the bipartition $(A \cap V_L, B \cap V_L)$ of V_L is also a bisection of $L(G)$. Consider the labels of the edges incident to r in G'. Observe that every two consecutive edges in the Eulerian tour except for the first and last edge have different labels. Moreover, the first edge has label a. If the last edge has label a, we have $|A| = |B| + 1$ and hence $|A \cap V_L| + 1 = |B \cap V_L|$. Otherwise, the last edge has label b, we have $|A| = |B|$ and hence $|A \cap V_L| = |B \cap V_L|$. Therefore, $(A \cap V_L, B \cap V_L)$ is a bisection of $L(G)$.

If G' has two or more connected components, we apply the same argument to each connected component and appropriately construct a bipartition of V_L. It is not hard to see that this bipartition also satisfies the condition in Lemma 2.

Since, given a line graph, we can compute its underlying graph [15,19] and an Eulerian tour in linear time, Max Bisection on line graphs can be solved in linear time.

6 Conclusion

In this paper, we show that there is an $O(2^t (tn)^2)$ time algorithm for Min Bisection and Max Bisection, provided that a width-t tree decomposition is given as input. This running time matches the conditional lower bound given by Eiben et al. [7] based on (min, +)-convolution. We also show that the exponential dependency of treewidth in our running time is asymptotically optimal under the Strong Exponential Time Hypothesis.

For unweighted graphs, Eiben et al. showed that the polynomial dependency can be slightly improved: They gave an $O(8^t t^{O(1)} n^{1.864} \log n)$-time algorithm for Min Bisection using an extension of the fast (min, +)-convolution technique due to Chan et al. [4]. It would be interesting to know whether a similar improvement can be applied to our case.

We also show that Min Bisection and Max Bisection are NP-hard even for several restricted graph classes. In particular, both problems are NP-hard even on unweighted bipartite graphs, which is in contrast with the tractability of MINCUT and MAXCUT on this graph classes. However, there are several open problems related to these results. One of the most notable open questions would be to reveal the complexity of MIN BISECTION on planar graphs.

References

1. Bodlaender, H.L., Hagerup, T.: Parallel algorithms with optimal speedup for bounded treewidth. SIAM J. Comput. **27**(6), 1725–1746 (1998)
2. Bodlaender, H.L., Jansen, K.: On the complexity of the maximum cut problem. Nordic J. Comput. **7**(1), 14–31 (2000)
3. Bui, T.N., Leighton, F.T., Chaudhuri, S., Sipser, M.: Graph bisection algorithms with good average case behavior. Combinatorica **7**(2), 171–191 (1987)
4. Chan, T.M., Lewenstein, M.: Clustered integer 3SUM via additive combinatorics. In: Proceedings of the Forty-Seventh Annual ACM Symposium on Theory of Computing, STOC 2015, pp. 31–40. Association for Computing Machinery, New York, NY, USA (2015)
5. Díaz, J., Kamiski, M.: Max-cut and max-bisection are np-hard on unit disk graphs. Theor. Comput. Sci. **377**(1–3), 271–276 (2007)
6. Díaz, J., Mertzios, G.B.: Minimum bisection is NP-hard on unit disk graphs. Inf. Comput. **256**, 83–92 (2017)
7. Eiben, E., Lokshtanov, D., Mouawad, A.E.: Bisection of bounded treewidth graphs by convolutions. In: Bender, M.A., Svensson, O., Herman, G. (eds.) 27th Annual European Symposium on Algorithms, ESA 2019. Leibniz International Proceedings in Informatics (LIPIcs), vol. 144, pp. 42:1–42:11. Schloss Dagstuhl-Leibniz-Zentrum fuer Informatik, Dagstuhl, Germany (2019)
8. Garey, M., Johnson, D., Stockmeyer, L.: Some simplified NP-complete graph problems. Theoret. Comput. Sci. **1**(3), 237–267 (1976)
9. Guruswami, V.: Maximum cut on line and total graphs. Discret. Appl. Math. **92**(2–3), 217–221 (1999)
10. Hadlock, F.: Finding a maximum cut of a planar graph in polynomial time. SIAM J. Comput. **4**(3), 221–225 (1975)
11. Impagliazzo, R., Paturi, R.: On the complexity of k-sat. J. Comput. Syst. Sci. **62**(2), 367–375 (2001)
12. Jansen, K., Karpinski, M., Lingas, A., Seidel, E.: Polynomial time approximation schemes for max-bisection on planar and geometric graphs. SIAM J. Comput. **35**(1), 110–10 (2005)
13. Karp, R.M.: Reducibility among combinatorial problems. In: Miller, R.E., Thatcher, J.W., Bohlinger, J.D. (eds.) Complexity of Computer Computations. The IBM Research Symposia Series, pp. 85–103. Springer, Boston (1972). https://doi.org/10.1007/978-1-4684-2001-2_9
14. Kloks, T. (ed.): Treewidth. LNCS, vol. 842. Springer, Heidelberg (1994). https://doi.org/10.1007/BFb0045375
15. Lehot, P.G.H.: An optimal algorithm to detect a line graph and output its root graph. J. ACM **21**(4), 569–575 (1974)
16. Lokshtanov, D., Marx, D., Saurabh, S.: Known algorithms on graphs of bounded treewidth are probably optimal. ACM Trans. Algorithms **14**(2), 1–30 (2018)
17. Orlova, G.I., Dorfman, Y.G.: Finding the maximal cut in a graph. Eng. Cybern. **10**, 502–506 (1972)

18. Pocai, R.V.: The complexity of simple max-cut on comparability graphs. Electron. Notes Discret. Math. **55**, 161–164 (2016)
19. Roussopoulos, N.D.: A max $\{m, n\}$ algorithm for determining the graph H from its line graph G. Inf. Process. Lett. **2**(4), 108–112 (1973)
20. Shih, W.K., Wu, S., Kuo, Y.S.: Unifying maximum cut and minimum cut of a planar graph. IEEE Trans. Comput. **39**(5), 694–697 (1990)

Influence Maximization Under the Non-progressive Linear Threshold Model

T.-H. Hubert Chan[1], Li Ning[2](✉), and Yong Zhang[2]

[1] Department of Computer Science, The University of Hong Kong, Hong Kong, China
hubert@cs.hku.hk
[2] Shenzhen Institutes of Advanced Technology, CAS, Shenzhen, China
{li.ning,zhangyong}@siat.ac.cn

Abstract. The *linear threshold model* [15] has been introduced to study the influence spreading behavior in the network, for instance, the spread of virus and innovations, and the objective of influence maximization is to choose a set of initially active nodes subject to some budget constraints, such that the expected number of active nodes over time is maximized.

In this paper, we extends the classic linear threshold model to capture the non-progressive behavior, i.e. the active nodes are allowed to become inactive again. The information maximization problem under our model is proved to be NP-Hard, even for the case when the underlying network has no directed cycles. The first result of this paper is negative. In general, the objective function of the extended linear threshold model is no longer submodular, and hence the *hill climbing* approach that is commonly used in the existing studies is not applicable. Next, as the main result of this paper, we prove that if the underlying information network is directed acyclic, the objective function is submodular (and monotone). Therefore, in directed acyclic networks with a specified budget we can achieve $\frac{1}{2}$-approximation on maximizing the number of active nodes over a certain period of time by a deterministic algorithm, and achieve the $(1 - \frac{1}{e})$-approximation by a randomized algorithm.

Keywords: Influence maximization · Non-progressive model · Linear threshold model

1 Introduction

We consider the problem of an advertiser promoting a product in a social network. The idea of *viral marketing* [7,13] is that with a limited budget, the advertiser can persuade only a subset of individuals to use the new product, perhaps by giving out a limited number of free samples. Then the popularity of the product is spread by word-of-mouth, i.e. through the existing connections between users in the underlying social network.

T.-H. Hubert Chan was partially supported by the Hong Kong RGC under the project 17200418.

Information networks have been used to model such cascading behavior [7, 11,14,15]. An information network is a directed edge-weighted graph, in which a node represents a user, whose behavior is influenced by its (outgoing) neighbors, and the weight of an edge reflects how influential the corresponding neighbor is. A node adopting the new behavior is *active* and is otherwise *inactive*. The *threshold model* [12,15] is one way to model the spread of the new behavior. The resistance of a node v to adopt the new behavior is represented by a random threshold θ_v (higher value means higher resistance), where the randomness is used to model the different susceptibility of different users. The new behavior is spread in the information network in discrete time steps. An inactive node changes its state to active if the weighted influence from the active neighbors in the previous time step reaches its threshold. We consider the *non-progressive* case where an active node could revert back to the inactive state if the influence from its neighbors drops below its threshold.

Our Contribution. The non-progressive linear threshold model proposed in this paper is a natural extension to the well known linear threshold model [15]. In Sect. 2, the model is formally defined, as well as the influence maximization problem. In most existing works, for a set of initially active nodes, the influence is measured by the maximum number of active nodes. Since the existing works consider the progressive case, hence the number of active nodes increases step by step and achieves the maximum after at most n time steps, where n is the number of the nodes. However, in the non-progressive case, it is possible that the active status never become stable. Hence, we introduce the average number of active nodes over a time period to measure the influence. Similar to the progressive case, the influence maximization problem considering the non-progressive linear threshold model is also NP-hard (Sect. 3). In order to approximate the optimal within a constant factor, a commonly used approach is to prove the monotone and submodular property of the objective function. Then the constant approximation ratio algorithms are promised by using the results of Fisher et al. [10] and Calinescu et al. [3]. However, this approach is not generally applicable for the non-progressive linear threshold model, since the average number of active nodes is possibly not submodular. As the main result of this paper, we studied the case when the information network is acyclic. As consistent with the intuition, the expected influence under the acyclic networks is submodular and hence Fisher's technique (and Calinescu's technique) is applicable to achieve the constant approximation. It should be noted that although the acyclic case looks much simpler than the general ones (where directed cycles may exist), the solution to maximize the expected influence is not that easy. As it is proved in Sect. 3, the problem of influence maximization is still NP-Hard even for the case under acyclic information networks. Futhermore, to prove the submodularity of the expected influence, we still need some tricky technique to handle complicated association between the status of the nodes. To see the association, consider time $t > 0$, the status of nodes at time t are associated if they share some common ancestors and the threshold of such an ancestor affects the status of its descendants. As this kind of association exist, it requires more carefully consideration of the nodes status and the analysis consequently become more complicate. In Sect. 4, we introduce an equivalent process (called Path Effect) for the non-progressive linear threshold model, and then the deep connection between this process and the random walk is proved via a coupling technique, which consequently leads to our final conclusion of the submodularity of the expected influence (under non-progressive linear threshold model).

Related Works. The cascading behavior in information networks was first studied in the computer science community by Kempe, Kleinberg and Tardos [15]. They considered the Independent Cascade Model and the Linear Threshold Model, the latter of which we generalize in this paper. Their main focus was the progressive case, and only reduced the non-progressive case to the progressive one by assigning a new independent random threshold to each node at every time step such that the resulting objective function is still submodular.

Kempe et al. [14,15] have also shown that the influence maximization problem in such models is NP-hard. Researchers often first show that the objective functions in question are submodular and then apply submodular function maximization methods to obtain constant approximation ratio. An example of such methods is the *Standard Greedy Algorithm*, which is analyzed by Nemhauser and Fisher et al. [10,18]. Loosely speaking, the Standard Greedy Algorithm (also known as the *Hill Climbing Algorithm*) starts with an empty solution, and in each iteration while there is still enough budget, we expand the current solution by including an additional node that causes the greatest increase in the objective function. Although the costs for transient and permanent nodes are different in our model, the budget constraint can still be described by a matroid. Under the matroid constraint, Fisher et al. [10] showed that the Standard Greedy Algorithm achieves $\frac{1}{2}$-approximation, and Calinescu et al. [3] introduced a randomized algorithm that achieves $(1 - \frac{1}{e})$-approximation in expectation.

In the above submodular function maximization algorithms, the objective function needs to be accessed in each iteration. However, to calculate the exact value of the objective function is in general hard [6]. One way to resolve this is to estimate the value of the objective function by sampling. Some works have used other ways to overcome this issue. To improve the efficiency of the Standard Greedy Algorithm, Leskovec et al. [16] showed a *Cost-Effective Lazy Forward* scheme, which makes use of the submodularity of the objective function and avoids the evaluation of influence on those nodes for which the incremental influence in the previous iteration is less than that of some already evaluated node in the current iteration. This scheme has been shown more efficient than the Standard Greedy Algorithm by experiments. Chen et al. [5] also designed an improved scheme to speed up the Standard Greedy Algorithm by using some efficiently computable heuristics that have similar performance.

Chen et al. [4] considered how *positive* and *negative* opinions spread in the same network, which can be interpreted as the influence process involving two agents. The influence maximization problem considering multiple competing agents in an information network has also been studied in [2,8]. We follow a similar setting in which a new comer can observe the strategies of existing agents, and stategizes accordingly to maximize his influence in the network.

Mossel and Roch [17] have shown that under more general submodular threshold functions (as opposed to linear threshold functions), the objective function is still submodular and hence the same maximization framework can still be applied.

Similar to our approach, the relationship between influence spreading and random walks has been investigated by Asavathiratham et al. [1], and Even-Dar and Shapira [9] in other information network models.

2   Preliminaries

Definition 1 (Information Network). *An information network is a directed weighted graph $G = (V, E, b)$ with node set V and edge set E, where each edge $(v, u) \in E$ has some positive weight $0 < b_{vu} < 1$ which intuitively represents the influencing power of u on v. Denote the set of outgoing neighbors of a node v by $\Gamma(v) := \{u \in V \mid (v, u) \in E\}$. In addition, for each $v \in V$, the total weight of its outgoing edges is at most 1, i.e. $\sum_{u \in \Gamma(v)} b_{vu} \leq 1$.*

Without loss of generality, we assume that in the considered information network G, $\sum_{u \in \Gamma(v)} b_{vu}$ is exactly 1 for every node $v \in V$. To achieve this requirement, for any given G, we add a void node d and for each node $v \neq d$, we include (v, d) into the set of edges, and set $b_{vd} := 1 - \sum_{u \in \Gamma(v) \setminus \{d\}} b_{vu}$. We can add a self loop with weight 1 at the void node d. Furthermore, d is never allowed to be active initially, and hence will never be active. Unless explicitly specified, when we use the term *node* in general, we mean a node other than the void node.

Next, we formally describe the extension of the classic linear threshold model for the adaption of the non-progressive behavior. A new feature of our model is that an initially active node can be either *transient* or *permanent*.

Model 1 (Non-progressive Linear Threshold Model (NLT)). *Consider an information network $G = (V, E, b)$. Each node in V is associated with a threshold θ_v, which is chosen from $(0, 1)$ independently and uniformly at random. At time $t \geq 0$, every node v is either* active $\mathcal{A}$ *or* inactive $\mathcal{N}$. *Denote the set of active nodes at time t by A_t. In the influence process, given a transient initial set $A \subseteq V$, a permanent initial set $\widehat{A} \subseteq V$, and a configuration of thresholds $\theta = \{\theta_v\}_{v \in V}$, the nodes update their status according to the following rules.*

1. *At time $t = 0$, $A_0 := A \cup \widehat{A}$.*
2. *At time $t > 0$, for each node $v \in V \setminus \widehat{A}$, compute the activation function $f_v(A_{t-1}) := \sum_{u \in A_{t-1} \cap \Gamma(v)} b_{vu}$. Then let $A_t := \{v \in V \setminus \widehat{A} \mid f_v(A_{t-1}) \geq \theta_v\} \cup \widehat{A}$.*

Without loss of generality, we can assume $A \cap \widehat{A} = \emptyset$, otherwise we can use $A \setminus \widehat{A}$ as the transient initial set instead. Given a transient initial set A and a permanent initial set $\widehat{A}$, we measure the influence of the agent by the average number of active nodes over T time steps, where T is some pre-specified time scope in which the process evolves. Observe that once the initial sets and the configuration of the thresholds are given, the active sets A_t's are totally determined.

Definition 2 (Influence Function and Expected Influence). *Given an information network G, a transient initial set A, a permanent initial set $\widehat{A}$, and a configuration θ of thresholds, the average influence over time period $[1,T]$ is defined as $\sigma_\theta^{[1,T]}(A, \widehat{A}) := \frac{1}{T}\sum_{t=1}^{T} |A_t|$. For simplicity, we ignore the super-script $[1,T]$ in σ when the target period is clear from the context. We define the expected influence $\overline{\sigma}$ as the expectation of $\sigma_\theta(A, \widehat{A})$ over the random choice of θ, i.e., $\overline{\sigma}(A, \widehat{A}) := \mathbb{E}_\theta[\sigma_\theta(A, \widehat{A})]$.*

Definition 3 (Influence Maximization Problem). *In an information net-work G, suppose the advertising cost of a transient initial node is c and that of a permanent initial node is $\widehat{c}$, where the costs are uniform over the nodes. Given a budget K, the goal is to find a transient initial set A and a permanent initial set $\widehat{A}$ with total cost $c \cdot |A| + \widehat{c} \cdot |\widehat{A}|$ at most K such that $\overline{\sigma}(A, \widehat{A})$ is maximized.*

The most technical part of the paper is to show that the objective function $\overline{\sigma}$ is submodular so that the maximization techniques of Fisher et al. [10] can be applied.

Definition 4 (Submodular, Monotone). *A function $f : 2^V \to \mathbb{R}$ is submod-ular if for any $A \subseteq B \subseteq V$ and $w \in V \backslash B$, $f(B \cup \{w\}) - f(B) \leq f(A \cup \{w\}) - f(A)$ holds. A function f is monotone if for any $A \subseteq B$, $f(A) \leq f(B)$. A function $g : 2^V \times 2^V \to \mathbb{R}$ is submodular (monotone), if keeping one argument constant, the function is submodular (monotone) as a function on the other argument.*

In order to facilitate the analysis of the influence process, we define indicator variables to consider the behavior of individual nodes at every time step.

Definition 5 (Indicator Variable). *In an information network G, given a transient initial set A and a permanent initial set $\widehat{A}$, a node v and a time t, let $X_v^t(A, \widehat{A})$ be the indicator random variable that takes value 1 if node v is active at time t, and 0 otherwise. When $\widehat{A} = \emptyset$, we sometimes write $X_v^t(A) := X_v^t(A, \emptyset)$.*

The indicator variable's usefulness is based on the following equality:

$$\overline{\sigma}(A, \widehat{A}) = \mathbb{E}\left[\frac{1}{T}\sum_{t=1}^{T} |A_t|\right] = \frac{1}{T}\sum_{t=1}^{T}\sum_{v \in V} \mathbb{E}\left[X_v^t(A, \widehat{A})\right].$$

Hence, if the function $(A, \widehat{A}) \mapsto \mathbb{E}[X_v^t(A, \widehat{A})]$ is submodular and monotone, then so is $\overline{\sigma}$.

3   Hardness of Maximization Problem

We outline an NP-hardness proof for the maximization problem in our setting via a reduction from vertex cover similar to that in [14,15]. We show that the problem is still NP-hard, even for the special case when the network is acyclic, and each transient node and each permanent has the same cost, which means only permanent nodes will be used.

Theorem 1. *The influence maximization problem under the non-progressive linear threshold model is NP-hard even when the network is a directed acyclic graph and all the initially active nodes are permanent.*

Proof. Given an undirected graph with n vertices, we pick an arbitrary linear ordering of the nodes and direct each edge accordingly to form a directed acyclic graph. We add a dummy node and for nodes with no outgoing edges, we add an edge from it to the dummy node. Hence, the network has $n + 1$ nodes. For each node, the weights of its outgoing edges are distributed uniformly. The number T of time steps under consideration is 1.

We claim that there is a vertex cover of size k for the constructed network *iff* there is a permanent initial set $\widehat{A}$ of size $k + 1$ such that $\overline{\sigma}(\emptyset, \widehat{A}) = n + 1$.

Suppose there is a vertex cover S of size k, then adding the dummy node to S to form $\widehat{A}$ as the permanent initial set, all nodes will be active in the next time step with probability 1, and so $\overline{\sigma}(\emptyset, \widehat{A}) = n + 1$. On the other hand, if the permanent initial set $\widehat{A}$ has size $k + 1$ and $\overline{\sigma}(\emptyset, \widehat{A}) = n + 1$, then the dummy node must be in $\widehat{A}$, and suppose S the set of non-dummy nodes in $\widehat{A}$. If S does not form a vertex cover for the given graph, then there exists some edge (u, v), where both nodes u and v are inactive initially, and hence the probability that u is active in the next time step is strictly smaller than 1. $\square$

4   Acyclic Information Networks

In this section, we consider the information networks without directed cycles. (As mentioned, we can show that there exists the case when the cycles are allowed and the expected influence is not submodular. The detailed argument are left to the full version of this work, as well as the proofs of Lemma 3-6 in this section). As we proved in Sect. 3, the influence maximization problem under **NLT** model is still NP-Hard even when the underlying network has no directed cycles.

Note that with the assumption of acyclic information networks, for any node $v \in V$ other than the void node, the set of its outgoing neighbors $\Gamma(v)$ has no directed path back to v.[1] Intuitively, during the influence procedure, v's choice of threshold θ_v can never affect the states of nodes in $\Gamma(v)$. To describe this fact formally, we introduce a random object on Ω, where Ω is the sample space, which is essentially the set of all possible configurations of the thresholds.

---

[1] Hence, the self-loop at the void node does not really interfere with the acyclic assumption.

Definition 6 (States of Nodes over Time). *Suppose Ω is the sample space, and W is a subset of nodes. Define a random object $\Pi_W : \Omega \to \{\mathcal{A}, \mathcal{N}\}^{|W| \times T}$, such that for $\omega \in \Omega$, $v \in W$ and $1 \leq t \leq T$, $\Pi_W(\omega)(v,t)$ indicates the state of node v at time t at the sample point ω.*

Lemma 1 (Independence). *Suppose $v \in V$ and $\eta \in [0,1]$, and let W be any subset of V with no directed path to v. Then, we have $Pr[\theta_v \leq \eta \mid \Pi_W] = \eta$, under **NLT** model.*

Proof. In **NLT** model, the randomness comes from the choices of thresholds $\theta = \{\theta_v\}_{v \in V}$. The sample space is actually the set of all possible configurations of thresholds θ.

Note that, the event $\theta_v \leq \eta$ is totally determined by the choice of θ_v, and the value of Π_W is totally determined by the choice of all θ_u's such that $u \neq v$. From the description of **NLT** model, the choices of thresholds are independent over different nodes. This implies that, for any Q in the range of Π_W, the events $\theta_v \leq \eta$ and $\Pi_W = Q$ are independent.

Hence, we get $Pr[\theta_v \leq \eta \mid \Pi_W] = Pr[\theta_v \leq \eta] = \eta$. $\qquad\square$

4.1 Connection to the Random Walk

Consider time $t > 0$. The status of nodes at time t may be associated, since they can share some common ancestors and the threshold of such an ancestor affects the status of its descendants. This association between different nodes cause more complicates for the analysis. In order to assist the analysis and handle the association carefully, we introduce a random walk process and show that this random walk process share an interesting connection with **NLT** model. Next, we introduce the random walk process.

Model 2 (Random Walk Process (RW)). *Consider an information network $G = (V, E, b)$. For any given node $v \in V$, we define a random walk process as follows.*

- *At time $t = 0$, the walk starts at node v.*
- *Suppose at some time t, the current node is u. A node $w \in \Gamma(u)$ is chosen with probability b_{uw}. The walk moves to node w at time $t + 1$.[2]*

Definition 7 (Reaching Event). *For any node v, subset $C \subseteq V$ and $t \geq 0$, we use $R_v^t(C)$ to denote the event that a random walk starting from v would reach a node in C at precisely time t.*

Next, we show the connection between the **NLT** model and the **RW** model for the case when the permanent initial set is empty. The more general case (arbitrary permanent initial set) will be considered later.

---

[2] Observe that if w is the void node, then the walk remains at w.

Lemma 2 (Connection between NLT model and RW model). *Consider an acyclic information network G, and let v be a non-void node, and $1 \leq t \leq T$. On the same network G, consider the* **NLT** *process on a transient initial set A and the* **RW** *process starting at v. Then, $\mathbb{E}[X_v^t(A)] = Pr[R_v^t(A)]$.*

Proof. This lemma is the key point in our argument, for which the proof is not obvious. To assist the proof of Lemma 2, we next introduce a process called "Path Effect" which is an "equivalent" presentation of the **NLT** model, and devote all the remaining part of this subsection to this lemma.   □

We next introduce the **PE** process which augments the **NLT** model and defines (random) auxiliary array structures P_v^t known as the *influence paths* to record the influence history. Intuitively, if v becomes active at time t, then the path P_v^t shows which of the initially active nodes is responsible. An important invariant is that node v is active at time t if and only if $P_v^t[0] \in A$.

Model 3 (Path Effect Process (PE)). *Consider an information network $G = (V, E, b)$. Each node $v \in V$ is associated with a threshold θ_v, which is chosen from $(0, 1)$ uniformly at random.*

Given a transient initial set $A \subseteq V$ and a configuration of the thresholds $\theta = \{\theta_v\}_{v \in V}$, for each node v and each time step, the influence paths P_v^t are constructed in the following influence procedure.

– *At time $t = 0$, for any $v \in V$, $P_v^0[0] = v$.*
– *At time $t > 0$, define the active set at the previous time step as $A_{t-1} = \{u \in V \mid P_u^{t-1}[0] \in A\}$. For each node $v \in V$, we compute $f_v(A_{t-1}) := \sum_{u \in \Gamma(v) \cap A_{t-1}} b_{vu}$. Then,*
   a. If $f_v(A_{t-1}) \geq \theta_v$, choose node $u \in \Gamma(v) \cap A_{t-1}$ with probability $\frac{b_{vu}}{f_v(A_{t-1})}$;
   b. If $f_v(A_{t-1}) < \theta_v$, choose node $u \in \Gamma(v) \setminus A_{t-1}$ with probability $\frac{b_{vu}}{1 - f_v(A_{t-1})}$.
   Once u is chosen, let $P_v^t[0, \ldots, t-1] := P_u^{t-1}[0, \ldots, t-1]$ and $P_v^t[t] = u$.[3]

Remark 1. Observe that given an information network with a transient initial set A and a configuration θ of thresholds. Both of the **NLT** model and the **PE** process produce exactly the same active set A_t at each time step t.

Definition 8 (Source Event). *Consider an information network G on which the* **PE** *process is run on the initial active set A. For any subset $C \subseteq V$, we use $I_v^t(C)$ to denote the event that $P_v^t[0]$ belongs to C. If $C \subseteq A$, this event means v's state at time t is the same as those of the nodes in A at time 0 and hence is active. We shall see later on that the event $I_v^t(C)$ is independent of A and hence the notation has no dependence on A.*

When the given network is acyclic, the **PE** process has an interesting property.

---

[3] Observe that if v is the void node, then $P_v^t[0, \ldots, t] = [void, \ldots, void]$.

Lemma 3 (Acyclicity Implies Independence of Choice). *Consider an information network $G = (V, E, b)$ on which the* **PE** *process is run with some initial active set. If G is acyclic, for any non-void node $v \in V$ and node $u \in \Gamma(v)$, we have $Pr[P_v^t[t] = u \mid \Pi_W] = b_{vu}$, where $W = \Gamma(v)$. Recall that Π_W carries the information about the states of the nodes in W at every time step.*

Recall the events $I_v^t(C)$ and $R_v^t(C)$ introduced in Definitions 8 and 7. The following lemma immediately implies Lemma 2 with $C = A$, and using the observation $\mathbb{E}[X_v^t(A)] = Pr[X_v^t(A) = 1] = Pr[I_v^t(A)]$ from Remark 1.

Lemma 4 (Connection between the PE process and the RW process). *Suppose the information network G is acyclic, and A is the transient initial set. For any $C \subseteq V$, any non-void node $v \in V$ and $t \geq 0$, we have $Pr[I_v^t(C)] = Pr[R_v^t(C)]$. In particular, the probability $Pr[I_v^t(C)]$ is independent of A.*

In the next subsection, we will reduce the general case with non-empty permanent initial set to the case when only transient initial set. Furthermore, we can prove the final conclusion (Theorem 3).

4.2   Submodularity of Acyclic NLT

At first, we consider the case where the permanent initial set $\widehat{A}$ is non-empty. We show that this general case can be reduced to the case where only transient initial set A is non-empty, by the following transformation. Suppose G is an information network, with transient initial set A and permanent initial set $\widehat{A}$, and T is the number of time steps to be considered. Consider the following transformation on the network instance. For each node $y \in \widehat{A}$, do the following:

1. Add a chain D_y of T dummy nodes to the network: starting from the head node of the chain, exactly one edge with weight 1 points to the next node, and so on, until the end node is reached.
2. Remove all outgoing edges from y. Add exactly one outgoing edge with weight 1 from y to the head of the chain D_y

See Fig. 1 for an example of the chain of dummy nodes. Let $D := \cup_{y \in \widehat{A}} D_y$ be the set of dummy nodes. We call the new network $\overline{G}(\widehat{A})$ the transformed network of G with respect to $\widehat{A}$. When there is no risk of confusion, we simply write $\overline{G}$. The transformed instance on $\overline{G}(\widehat{A})$ only has $A \cup \widehat{A} \cup D$ as the transient initial set and no permanent initial node. The initially active dummy nodes in D ensure that every node $y \in \widehat{A}$ is active for T time steps. We use the notation convention that we add an overline to a variable (e.g., $\overline{X}$), if it is associated with the transformed network.

Remark 2. For any non-dummy, non-void node v,

$$X_v^t(A, \widehat{A}) = \overline{X}_v^t(A \cup \widehat{A} \cup D).$$

Fig. 1. The chain of dummy nodes

Lemma 5. *Suppose we are given an instance on information network G, with transient initial set A and permanent initial set $\widehat{A}$. Let v be any non-void node in G and $0 \leq t \leq T$. Suppose in the transformed network $\overline{G}(\widehat{A})$, for any subset C of nodes in G, $\overline{R}_v^t(C)$ is the event that starting at v, the **RW** process on $\overline{G}$ for t steps ends at a node in C. Then,*

$$\mathbb{E}[X_v^t(A, \widehat{A})] = \sum_{u \in A} Pr[\overline{R}_v^t(\{u\})] + \sum_{y \in \widehat{A}} \sum_{i=0}^{t} Pr[\overline{R}_v^i(\{y\})].$$

Definition 9 (Passing-Through Event). *Let G be an information network. For any node v, subset $C \subseteq V$ and $t \geq 0$, we use $S_v^t(C)$ to denote the event that a **RW** process on G starting from v would reach a node in C at time t **or before**.*

Lemma 6 (General Connection between NLT and RW). *Suppose G is an acyclic information network, and let v be a non-void node, and $1 \leq t \leq T$. On the same network G, consider the **NLT** model with transient initial set A and permanent initial set $\widehat{A}$, and the **RW** process starting at v. Then, $\mathbb{E}[X_v^t(A, \widehat{A})] = Pr[R_v^t(A) \cup S_v^t(\widehat{A})].$*

Theorem 2. (Submodularity and Monotonicity of $\mathbb{E}[X_v^t(A, \widehat{A})]$). *Consider the **NLT** model on an acyclic information network G with transient initial set A and permanent initial set $\widehat{A}$. Then, the function $(A, \widehat{A}) \mapsto \mathbb{E}[X_v^t(A, \widehat{A})]$ is submodular and monotone.*

Proof. For notational convenience, we drop the superscript t and the subscript v, and write for instance $X(A, \widehat{A}) := X_v^t(A, \widehat{A})$. For the reaching and the passing-through events associated with the Random Walk Process in G, we write $R(A) := R_v^t(A)$ and $S(A) := S_v^t(A)$

It is sufficient to prove that, for any $A \subseteq B \subseteq V$, $\widehat{A} \subseteq \widehat{B} \subseteq V$, and node $w \notin (B \cup \widehat{B})$, the following inequalities hold:

$$\mathbb{E}[X(A \cup \{w\}, \widehat{A})] - \mathbb{E}[X(A, \widehat{A})] \geq \mathbb{E}[X(B \cup \{w\}, \widehat{B})] - \mathbb{E}[X(B, \widehat{B})]; \quad (1)$$
$$\mathbb{E}[X(A, \widehat{A} \cup \{w\})] - \mathbb{E}[X(A, \widehat{A})] \geq \mathbb{E}[X(B, \widehat{B} \cup \{w\})] - \mathbb{E}[X(B, \widehat{B})]. \quad (2)$$

By Lemma 6, for any subsets C and $\widehat{C}$ such that $w \notin (C \cup \widehat{C})$, $x(C \cup \{w\}, \widehat{C}) - x(C, \widehat{C}) = Pr[R(C \cup \{w\}) \cup S(\widehat{C})] - Pr[R(C) \cup S(\widehat{C})] = Pr[R(\{w\}) \setminus S(\widehat{C})]$, where the last equality follows from definitions of reaching and passing-through events. Hence, inequality (1) follows because $\widehat{A} \subseteq \widehat{B}$ implies that $R(\{w\}) \setminus S(\widehat{A}) \supseteq R(\{w\}) \setminus S(\widehat{B})$.

Similarly, $\mathbb{E}[X(C,\widehat{C} \cup \{w\})] - \mathbb{E}[X(C,\widehat{C})] = Pr[S(\{w\}) \setminus (R(C) \cup S(\widehat{C}))]$. Hence, inequality (2) follows because $R(A) \cup S(\widehat{A}) \subseteq R(B) \cup S(\widehat{B})$.   □

Corollary 1 (Objective Function is Submodular and Monotone). *With the same hypothesis as in Theorem 2, the function $(A, \widehat{A}) \mapsto \overline{\sigma}(A, \widehat{A})$ is submodular and monotone.*

At the end, we achieve the main result of this paper.

Theorem 3. *Given an acyclic information network, a time period $[1, T]$, a budget K and advertising costs (transient or permanent) that are uniform over the nodes, an advertiser can use the Standard Greedy Algorithm to compute a transient initial set A and a permanent initial set $\widehat{A}$ with total cost at most K in polynomial time such that $\overline{\sigma}(A, \widehat{A})$ is at least $\frac{1}{2}$ of the optimal value. Moreover, there is a randomized algorithm that outputs A and $\widehat{A}$ such that the expected value (over the randomness of the randomized algorithm) of $\overline{\sigma}(A, \widehat{A})$ is at least $1 - \frac{1}{e}$ of the optimal value, where e is the natural number.*

Proof. We describe how Theorem 3 is derived. Recall that the advertiser is given a budget K, and the cost per transient node is c and the cost per permanent node is $\widehat{c}$. Observe that if the advertiser uses k transient nodes, where $k \leq \lfloor \frac{K}{c} \rfloor$, then there can be at most $\widehat{k} := \lfloor \frac{K-kc}{\widehat{c}} \rfloor$ permanent nodes. Hence, for each such guess of k and the corresponding $\widehat{k}$, the advertiser just needs to consider the maximization of the submodular and monotone function $(A, \widehat{A}) \mapsto \overline{\sigma}(A, \widehat{A})$ on the matroid $\{(A, \widehat{A}) : |A| \leq k, |\widehat{A}| \leq \widehat{k}\}$, for which $\frac{1}{2}$-approximation can be obtained in polynomial time using the techniques of Fisher et al. [10]. A randomized algorithm given by Calinescu et al. [3] achieves $(1-\frac{1}{e})$-approximation in expectation.   □

5   Conclusions

In this work, we extended the classic linear threshold model of the influence process in the networks, to allow the non-progressive behavior. Based on the reduction from the vertex cover problem, we have proved that under the proposed non-progressive linear threshold model, it is hard to maximize the average number of active nodes by initially activating someones with a budge constraint. Although in general the expected influence is not submodular, it was shown monotone and submodular when the underlying network is acyclic, in which case, the problem of influence maximization is still NP-Hard, and thus the hill climbing approach can be applied to achieve a constant approximation to the optimum.

References

1. Asavathiratham, C., Roy, S., Lesieutre, B., Verghese, G.: The influence model. Control Syst. IEEE **21**(6), 52–64 (2001)

2. Bharathi, S., Kempe, D., Salek, M.: Competitive influence maximization in social networks. In: Proceedings of the 3rd International Conference on Internet and Network Economics, pp. 306–311 (2007)
3. Călinescu, G., Chekuri, C., Pál, M., Vondrák, J.: Maximizing a monotone submodular function subject to a matroid constraint. SIAM J. Comput. **40**(6), 1740–1766 (2011)
4. Chen, W., et al.: Influence maximization in social networks when negative opinions may emerge and propagate. In: SDM, pp. 379–390 (2011)
5. Chen, W., Wang, Y., Yang, S.: Efficient influence maximization in social networks. In: Proceedings of the 15th ACM SIGKDD International Conference on Knowledge Discovery and Data Mining, pp. 199–208 (2009)
6. Chen, W., Yuan, Y., Zhang, L.: Scalable influence maximization in social networks under the linear threshold model. In: Proceedings of the 2010 IEEE International Conference on Data Mining, pp. 88–97 (2010)
7. Domingos, P., Richardson, M.: Mining the network value of customers. In: Proceedings of the 7th ACM SIGKDD International Conference on Knowledge Discovery and Data Mining, pp. 57–66 (2001)
8. Dubey, P., Garg, R., Meyer, B.D.: Competing for customers in a social network. Department of Economics Working Papers 06–01, Stony Brook University, Department of Economics (2006)
9. Even-Dar, E., Shapira, A.: A note on maximizing the spread of influence in social networks. In: Deng, X., Graham, F.C. (eds.) WINE 2007. LNCS, vol. 4858, pp. 281–286. Springer, Heidelberg (2007). https://doi.org/10.1007/978-3-540-77105-0_27
10. Fisher, M.L., Nemhauser, G.L., Wolsey, L.A.: An analysis of approximations for maximizing submodular set functions. II. In: Balinski, M.L., Hoffman, A.J. (eds.) Polyhedral Combinatorics. Mathematical Programming Studies, vol. 8, pp. 73–87. Springer, Berlin, Heidelberg (1978). https://doi.org/10.1007/BFb0121195
11. Goldenberg, J., Libai, B., Muller, E.: Using complex systems analysis to advance marketing theory development. Acad. Mark. Sci. Rev. **9**(3), 1–18 (2001)
12. Granovetter, M.: Threshold models of collective behavior. Am. J. Sociol. **83**(6), 1420–1443 (1978)
13. Jurvetson, S.: What exactly is viral marketing? Red Herring **78**, 110–112 (2000)
14. Kempe, D., Kleinberg, J., Tardos, E.: Influential nodes in a diffusion model for social networks. In: Proceedings of the 32nd International Colloquium on Automata, Languages and Programming, pp. 1127–1138 (2005)
15. Kempe, D., Kleinberg, J., Tardos, É.: Maximizing the spread of influence through a social network. In: Proceedings of the Ninth ACM SIGKDD International Conference on Knowledge Discovery and Data Mining, pp. 137–146. ACM (2003)
16. Leskovec, J., Krause, A., Guestrin, C., Faloutsos, C., VanBriesen, J., Glance, N.: Cost-effective outbreak detection in networks. In: Proceedings of the 13th ACM SIGKDD International Conference on Knowledge Discovery and Data Mining, pp. 420–429 (2007)
17. Mossel, E., Roch, S.: On the submodularity of influence in social networks. In: Proceedings of the 39th Annual ACM Symposium on Theory of Computing, pp. 128–134 (2007)
18. Nemhauser, G.L., Wolsey, L.A., Fisher, M.L.: An analysis of approximations for maximizing submodular set functions - I. Math. Program. **14**(1), 265–294 (1978)

Car-Sharing: Online Scheduling k Cars Between Two Locations

Songhua Li$^{(\boxtimes)}$, Leqian Zheng, and Victor C. S. Lee

Department of Computer Science, City University of Hong Kong,
Kowloon, Hong Kong SAR, China
{songhuali3-c,leqizheng2-c}@my.cityu.edu.hk, csvlee@cityu.edu.hk

Abstract. This paper studies a special setting of the online order-dispatching problem in the car-sharing system (CSS). Once an order is released online by a user at its booking time, it specifies the departure time and two locations (in which one is the departure location while the other is the destination). The CSS operator needs to determine whether to accept or reject an online order at its booking time. Once accepting an order, the operator should assign a car among k cars to serve the order right at its departure time, gaining a constant revenue paid by the user. However, a movement without taking an order (i.e., empty movement) incurs a constant cost for the operator. The goal is to schedule the k cars to serve a set of online released orders that maximizes the overall profits (i.e., the overall revenue minus the overall cost generated by accepted orders). With regard to the cost of empty movement and the time gap between an order's booking time and departure time, we consider different settings. We show the lower bounds on the competitive ratio for different settings respectively, and propose two online algorithms GD and BD, with their competitive ratios matching and approaching the lower bounds in corresponding settings respectively.

Keywords: Car-sharing system · Online algorithm · Online scheduling

1 Introduction

Over the past few years, people have witnessed a rapid growth in applications of the shared mobility [1], which provides effective solutions in relieving traffic congestion and saving operation costs. One popular mobility in the shared community is the car-sharing system (CSS, hereafter), of which the representatives include the Uber and the DiDi Express. In CSS, passengers submit their trip orders online via mobile applications. Normally, each online order, which is submitted at the booking time, consists of its departure time, its departure location, and its destination. The operator must determine whether to accept or reject an online order right at the booking time and serve each of its accepted orders right at the departure time. Hence, how to accept and serve those online orders effectively and efficiently is one of the central issues for the CSS operators. There are two major threads of relevant research.

© Springer Nature Switzerland AG 2020
M. Li (Ed.): FAW 2020, LNCS 12340, pp. 49–61, 2020.
https://doi.org/10.1007/978-3-030-59901-0_5

1.1 Related Work

The first thread is on the online bipartite matching in which the right- and left-side vertices in a bipartite graph represent the trip orders and the cars, respectively. Huang et al. [8,9] studied a fully online model of maximum cardinality matching where all vertices arrive online, which fits into the ride-sharing system (RSS) since both trip orders and drivers may appear online in RSS. They showed that the famous algorithm Ranking is 0.5211-competitive in the fully online model, while no online algorithm can beat a competitive ratio of 0.6317. Dickerson et al. [5] studied online matching with reusable resources under known adversarial distributions, where matched resources can become available again shortly. Zhao et al. [6] studied the online stable matching to solve the preference-aware task assignment in on-demand taxi dispatching systems, aiming to optimize the expected total profits and the satisfaction among workers and users. Bei et al. [7] proposed a 2.5-approximation algorithm to assign exactly $2n$ orders to n drivers, which is based on a two-phase perfect matching approach.

The second thread is about online scheduling problems. Luo et al. [2] studied a setting where there is only one car to serve online orders between two locations. In their setting, each served order yields a constant revenue r to the system, an empty movement (i.e., moving without taking an order) incurs a constant cost c ($0 \leqslant c \leqslant r$), and the objective is to accept a set of orders that maximizes the total profits. With regard to the time interval between an order's booking time and departure time, they considered two scenarios including the fixed time interval and the variable time interval; they showed lower bounds on the competitive ratio for both two scenarios and proposed a greedy algorithm with the competitive ratio matching the lower bounds respectively. Later, they extended the model to two cars [4] and showed that 2 is a tight lower bound on competitive ratio for the model with two cars, by proposing a 2-competitive algorithm. They [3] also studied the problem with k cars to serve online orders between two locations, where the departure time of orders is assumed to be a multiple of the travel time between the two locations and the cost for empty movement (i.e., moving while not serving an order) is removed. They showed tight lower bounds for some special scenarios by proposing a smart greedy algorithm.

1.2 Problem Description and Preliminaries

This paper studies the problem of online scheduling k cars to serve trip orders between two locations, denoted by 0 and 1 respectively, which inherits a basic model from [3]. Specifically, all the k cars are located at location 0 initially at time 0. The travel time between the two locations is a constant t. Each order contributes a constant revenue γ to the system once it is accepted and served by a car. An empty car at a location of $\{0,1\}$ may be dispatched to the other location to pick up a previously accepted order. Such an empty movement incurs a constant cost ς to the system. Note that any unprompted empty movement (i.e., an empty movement of a car that is not for serving an accepted order which departs from the opposite location) is not allowed in the setting. Denote $r_i =$

$(\widetilde{t_{r_i}}, t_{r_i}, p_{r_i})$ as the ith released order by users, in which $\widetilde{t_{r_i}}$, t_{r_i} and $p_{r_i} \in \{0, 1\}$ indicate the booking time, the departure time and the departure location (while the other location $(1 - p_{r_i})$ is the destination) of the order respectively. Denote R as the online sequence of orders released to the system, in which each is released at its booking time. Meanwhile, the operator must determine whether to accept an order or not immediately at its booking time. Once accepting an order r_i, the system must assign a car to arrive at the departure location p_{r_i} of the order right at the departure time t_{r_i} to serve the order (which implies the car reaches location $1 - p_{r_i}$ at time $t_{r_i} + t$). Each served order r_i contributes a profit $\gamma - \varsigma$ (resp. γ) to the system if its assigned car departs from $1 - p_{r_i}$ (resp. p_{r_i}) before t_{r_i} to pick-up r_i, and the order is called *with-cost* (resp. *without-cost*) served. We say two orders r_i and r_j are *in conflict* if they cannot be served by the same car[3], i.e., r_i and r_j are in conflict if and only if $|t_{r_i} - t_{r_j}| < t$ when $p_{r_i} = 1 - p_{r_j}$ or $|t_{r_i} - t_{r_j}| < 2t$ when $p_{r_i} = p_{r_j}$. The **objective** is to maximize the overall profit of the system that is achieved by accepting and serving a set of online released orders.

We distinguish our model from the model in [3] in two aspects: First, the departure time of an order can be any time point of a continuous timeline in our model, while it must be a multiple of the travel time t between 0 and 1 in [3]; Second, the cost ς for an empty movement in our model is not considered in the model of [3]. In addition, we inherit the terms kS2L-F and kS2L-V from [3], to specify the variants where the gap $t_{r_i} - \widetilde{t_{r_i}}$ between the departure time and the release time of an order r_i is a constant a $(\geqslant 0)$ and within a range $[b_l, b_u]$ respectively. To this end, we aim at online algorithms with provable good performance in terms of the competitive ratio [10] which is defined as follows. Given a sequence R of online orders of the problem, denote by I_{ALG} and I_{OPT} as the overall profit earned by an online algorithm ALG and an optimal algorithm (OPT) of the offline problem (which has the complete information of all orders in advance) respectively. The competitive ratio of an online algorithm (ALG) is defined as $\rho_{\text{ALG}} = \sup_R \frac{I_{\text{OPT}}}{I_{\text{ALG}}}$. Particularly, we say λ $(\geqslant 1)$ is a lower bound on the competitive ratio of the problem if $\lambda \leqslant \rho_{\text{ALG}}$ holds for any ALG.

1.3   Main Results

With regard to the parameters a, t, ς and γ of the model, we investigate different settings in both kS2L-F and kS2L-V variants respectively. We first present lower bounds on the competitive ratio for the settings respectively. Then, we show that a simple greedy algorithm can easily achieve the optimal worst-case performance, with its competitive ratio matches the lower bounds in most of the settings. For the only setting KS2L-F with $t \leqslant b_l$, in which the simple greedy algorithm does not perform well, we propose another algorithm (BD) with its competitive ratio proved to be close to the lower bound of the setting. Main results of this paper are summarized in Table 1, while the state-of-the-art results in the literature are summarized in Table 2. Due to space limit, some proofs are omitted.

Table 1. Our results: lower bounds and upper bounds (for $k \in \mathbb{N}$).

Problem	Booking constraints	LB ($0 \leqslant \varsigma < \gamma$)	UB ($0 \leqslant \varsigma < \gamma$)	LB ($\varsigma = \gamma$)	UB ($\varsigma = \gamma$)
kS2L-F	$0 \leqslant a < t$	1	1	1	1
kS2L-F	$t \leqslant a$	$\min\{\frac{k}{\lfloor\frac{2k\gamma}{3\gamma+\varsigma}\rfloor},$ $\frac{2k\gamma}{2k\gamma-(\gamma+\varsigma)\lceil\frac{2k\gamma}{3\gamma+\varsigma}\rceil}\}$	$\max\{\frac{k}{\lceil\frac{k\gamma}{2\gamma+\varsigma}\rceil},$ $\frac{2k\gamma}{k\gamma-\lceil\frac{k\gamma}{2\gamma+\varsigma}\rceil\varsigma}\}$	1	1
kS2L-V	$0 < b_u < t$	3	3	3	3
kS2L-V	$t \leqslant b_u$	$\min\{\frac{2k}{\lfloor\frac{2k\gamma}{3\gamma+\varsigma}\rfloor},$ $\frac{k(3\gamma-\varsigma)}{k\gamma-\varsigma\lceil\frac{2k\gamma}{3\gamma+\varsigma}\rceil}\}$	$\lceil\frac{k}{4}\rceil$ $(k \geqslant 4)$	$\min\{\frac{2k}{\lfloor\frac{k}{2}\rfloor},$ $\frac{2k}{k-\lceil\frac{k}{2}\rceil}\}$	$\lceil\frac{k}{4}\rceil$ $(k \geqslant 4)$

Table 2. The state-of-the-art results.

Problem	Booking constraints	LB ($0 \leqslant \varsigma < \gamma$)	UB ($0 \leqslant \varsigma < \gamma$)	LB ($\varsigma = \gamma$)	UB ($\varsigma = \gamma$)
kS2L-F	$0 \leqslant a < t$	1 (for $k = 1$), [2]	1 (for $k = 1$), [2]	1 (for $k = 1$), [2]	1 (for $k = 1$), [2]
kS2L-F	$t \leqslant a$	1.5, [3]	1.5 (for $k = 3i$, $i \in \mathbb{N}$), [3]	1 (for $k = 1$), [2]	1 (for $k = 1$), [2]
kS2L-V	$0 < b_u < t$	3 (for $k = 1$), [2]	3 (for $k = 1$), [2]	3 (for $k = 1$), [2]	3 (for $k = 1$), [2]
kS2L-V	$t \leqslant b_u$	$\frac{3\gamma-\varsigma}{\gamma-\varsigma}$ (for $k = 1$), [2]	$\frac{3\gamma-\varsigma}{\gamma-\varsigma}$ (for $k = 1$), [2]	$1 + 2\lceil\frac{b_u-b_l}{2t}\rceil$ (for $k = 1$), [2]	$1 + 2\lceil\frac{b_u-b_l}{2t}\rceil$ (for $k = 1$), [2]

2   Lower Bounds

The following lower bounds are constructed similarly. For the sake of space restriction, we only include the proof of Theorem 2 in this version.

Theorem 1. *For kS2L-F with $\varsigma \geqslant \gamma$ or $0 \leqslant a < t$, no deterministic online algorithm can achieve a competitive ratio better than 1.*

Theorem 2. *For kS2L-F with $t \leqslant a$ and $0 \leqslant \varsigma < \gamma$, no deterministic online algorithm can achieve a competitive ratio better than* $\min\{\frac{k}{\lfloor\frac{2k\gamma}{3\gamma+\varsigma}\rfloor},$ $\frac{2k\gamma}{2k\gamma-(\gamma+\varsigma)\lceil\frac{2k\gamma}{3\gamma+\varsigma}\rceil}\}$.

Proof. The adversary first releases a sequence $R_1 = \{r_i = (t, t + a, 1)|i \in \{1, 2, 3, ..., k\}\}$ of k orders successively. We discuss two cases.

Case 1. ALG does not accept any order in R_1. Later, the adversary does not release new order any more, and OPT accepts all orders in R_1. Hence, we have $\rho_{ALG} = \frac{I_{OPT}}{I_{ALG}} = +\infty$ since OPT can accept all the k orders in R_1 gaining a profit of $k(\gamma - \varsigma)$ while ALG does not achieve any profit.

Case 2. ALG accepts β orders in R_1. We further discuss two cases.

Case 2.1. $\beta \leqslant \lfloor\frac{2k\gamma}{3\gamma+\varsigma}\rfloor$. Later, the adversary does not release orders. We have $I_{OPT} = k(\gamma - \varsigma)$ and $I_{ALG} = \beta(\gamma - \varsigma)$. Hence, $\rho_{ALG} = \frac{I_{OPT}}{I_{ALG}} \geqslant \frac{k}{\beta}$ since OPT can accept all orders in R_1.

Algorithm 1. Greedy Dispatching (GD).

Input: k cars, online sequence $R = \{r_1, r_2 ..., r_n\}$ of orders;
Output: a set of accepted orders;

1: **while** a new order $r_j = (\widetilde{t_{r_j}}, t_{r_j}, p_{r_j})$ is released **do**
2:    **if** $\Psi(r_i) \neq \varnothing$ **and** $\max\limits_{x \in \Psi(r_i)} \{I_{GD}(B(x) \cup \{r_i\}) - I_{GD}(B(x))\} > 0$ **then**
3:       accept r_i, dispatch the car arg $\max\limits_{x \in \Psi(r_i)} \{I_{GD}(B(x) \cup \{r_i\}) - I_{GD}(B(x))\}$ to r_i
4:    **else**
5:       reject r_i;
6:    **end if**
7: **end while**

Case 2.2. $\beta \geqslant \lceil \frac{2k\gamma}{3\gamma+\varsigma} \rceil$. The adversary further releases another two sequences of orders $R_2 = \{r_{k+i} = (t + \varepsilon, t + a + \varepsilon, 0) | i \in \{1, 2, 3, ..., k\}\}$ and $R_3 = \{r_{2k+i} = (2t + \varepsilon, 2t + a + \varepsilon, 1) | i \in \{1, 2, ..., k\}\}$, where $0 < \varepsilon < a$. Thus, OPT can achieve a profit of $I_{OPT} = 2k\gamma$ by accepting all orders in R_2 and R_3. Meanwhile, $I_{ALG} = \beta(\gamma - \varsigma) + (k - \beta)2\gamma$. Hence, $\rho_{ALG} = \frac{I_{OPT}}{I_{ALG}} \geqslant \frac{2k\gamma}{2k\gamma - (\gamma + \varsigma)\beta}$.

Therefore, we have $\rho_{ALG} \geqslant \min\{\frac{k}{\lfloor \frac{2k\gamma}{3\gamma+\varsigma} \rfloor}, \frac{2k\gamma}{2k\gamma - (\gamma+\varsigma)\lceil \frac{2k\gamma}{3\gamma+\varsigma} \rceil}\}$.

Theorem 3. *For kS2L-V with $0 < b_u < t$, no deterministic online algorithm can achieve a competitive ratio better than 3.*

Theorem 4. *For kS2L-V with $t \leqslant b_u$ and $0 \leqslant \varsigma < \gamma$, no deterministic online algorithm can achieve a competitive ratio better than $\min\{\frac{2k}{\lfloor \frac{2k\gamma}{3\gamma+\varsigma} \rfloor}, \frac{3k\gamma - k\varsigma}{k\gamma - \varsigma \lceil \frac{2k\gamma}{3\gamma+\varsigma} \rceil}\}$.*

Theorem 5. *For kS2L-V with $t \leqslant b_u$ and $\varsigma = \gamma$, no deterministic online algorithm can achieve a competitive ratio better than $\min\{\frac{2k}{\lfloor \frac{k}{2} \rfloor}, \frac{2k}{k - \lceil \frac{k}{2} \rceil}\}$.*

3 Upper Bounds

3.1 Greedy Dispatching (GD) Algorithm

First, we present the Greedy Dispatching (GD) algorithm. For ease of expression, we refer to an order that departs from location 0 (resp. 1) as a $\overrightarrow{01}$ order (resp. $\overrightarrow{10}$ order). Denote B as the set of orders that are accepted by GD algorithm, and further, denote B^0 and B^1 as those $\overrightarrow{01}$ and $\overrightarrow{10}$ orders in B, respectively. Let $B(x)$ denote the set of orders assigned to the car x in GD, and $I_{GD}(B(x))$ denote the profit achieved by orders in $B(x)$ that are served by the car x. Denote

$$\Psi(r_i) = \{x \in \{1, \cdots, k\} | \forall r_j \in B(x), \begin{cases} |t_{r_i} - t_{r_j}| \geqslant 2t, & \text{when } p_{r_i} = p_{r_j} \\ |t_{r_i} - t_{r_j}| \geqslant t, & \text{when } p_{r_i} \neq p_{r'_j} \end{cases}\} \text{ as the}$$

set of cars in GD that are available to a new order r_i at its release time $\widetilde{t_{r_i}}$. We say a new order r_i is *acceptable* iff $\Psi(r_i) \neq \varnothing$.

GD (Algorithm 1) Description. A new order r_i is accepted by GD only when the following two *acceptable conditions* hold together: (i) $\Psi(r_i) \neq \varnothing$, which guarantees that r_i is acceptable by at least one car in GD; (ii) $\max\limits_{x \in \Psi(r_i)} I_{GD}(B(x) \cup \{r_i\}) - I_{GD}(B(x)) > 0$, i.e., the car can earn a non-zero profit from r_i.

For kS2L-F with either $\gamma = \varsigma$ or $0 \leqslant a < t$, and kS2L-V with $b_u < t$, we note that neither OPT nor GD can with-cost serve an order respectively, because a car cannot catch up an opposite order from the order's release time. Thus, we have the following Theorem 6 and 7, in which proofs are omitted due to space limitation.

Theorem 6. *GD is 1-competitive for kS2L-F with $\gamma = \varsigma$ or $0 \leqslant a < t$.*

Theorem 7. *For the kS2L-V with $0 < b_u < t$, GD is 3-competitive.*

Unlike in $a < t$, both GD and OPT possibly with-cost serve orders in $t \leqslant a$, respectively, since a car becomes available to catch up an opposite order from the order's release time. Thus, one can derive that the competitive ratio of GD is no better than $\frac{2\gamma}{\gamma - \varsigma}$ which becomes far larger than the lower bound $\min\{\frac{k}{\lfloor \frac{2k\gamma}{3\gamma + \varsigma} \rfloor},$ $\frac{2k\gamma}{2k\gamma - (\gamma + \varsigma)\lceil \frac{2k\gamma}{3\gamma + \varsigma} \rceil}\}$ of the setting when ς approaches to γ. To this end, we extend the GD algorithm to the GD_θ algorithm by including a new *acceptable condition* (iii) in the Line 2 of Algorithm 1: given a new order r_i and the set B of currently accepted orders by GD_θ, its indicator function $1_{(r_i,B)}$ should equal to 1, which is as defined below. Namely, GD_θ accepts a new order r_i only when r_i satisfies conditions (i)–(iii) together.

Definition of the Indicator Function $1_{(r_i,B)}$. GD_θ always accepts the first released order $r_1 \in R$, i.e., $1_{(r_1,B)} = 1$. Based on the departure time t_{r_1} of r_1, GD_θ divides the time line into continuous time frames $F_m = [2tm + t_{r_1}, 2tm + 2t + t_{r_1})$, for each $m \in \{0, 1, \cdots, \}$. Later, GD_θ always balances the number of accepted orders that depart from two different locations in each time frame respectively. More specific, GD_θ accepts at most $\theta = \lfloor \frac{k\gamma}{2\gamma + \varsigma} \rfloor$ $\overrightarrow{10}$ orders and at most $k - \theta$ (i.e., $k - \lfloor \frac{k\gamma}{2\gamma + \varsigma} \rfloor$) $\overrightarrow{01}$ orders in each time frame. When a new order $r_i = (\widetilde{t_{r_i}}, t_{r_i}, p_{t_{r_i}})$ is released, the indicator function of $1_{(r_i,B)}$ is defined by the following (1).

$$1_{(r_i,B)} = \begin{cases} 1, & p_{t_{r_i}} = 1 \text{ and } |\{r_b \in B^1 | t_{r_b} \in F_m\}| \leqslant \lfloor \frac{k\gamma}{2\gamma + \varsigma} \rfloor - 1 \\ 1, & p_{t_{r_i}} = 0 \text{ and } |\{r_b \in B^0 | t_{r_b} \in F_m\}| \leqslant k - \lfloor \frac{k\gamma}{2\gamma + \varsigma} \rfloor - 1 \quad (1) \\ 0, & \text{else} \end{cases}$$

Theorem 8. *GD_θ is $\max\{\frac{k}{\lfloor \frac{k\gamma}{2\gamma + \varsigma} \rfloor}, \frac{2k\gamma}{k\gamma - \lfloor \frac{k\gamma}{2\gamma + \varsigma} \rfloor \varsigma}\}$-competitive for the kS2L-F with $t \leqslant a$ and $0 \leqslant \varsigma < \gamma$.*

3.2   Balanced Dispatching (BD)

Next, we propose a new online algorithm **B**alanced **D**ispatching (BD) for the kS2L-V with $b_l \geqslant t$ (i.e., $t_r - \widetilde{t_r} \geqslant t$).

In BD, we divide the continuous time line into time slots $T_i = [(i-1)\cdot t, i\cdot t)$, where $i \in \mathbb{N}^+$. Denote B_i^0 (resp. B_i^1) as the set of accepted orders by BD that depart from location 0 (resp. 1) within time slot T_i. Accordingly, $|B_i^0|$ (resp. $|B_i^1|$) denotes the number of accepted orders by BD that depart from location 0 (resp. 1) within time slot T_i, and $B_i = B_i^0 \cup B_i^1$ denotes the accepted orders by BD in T_i, and $B = \bigcup_i B_i$ further denotes the set of accepted orders by BD. When a new order $r_j = (\widetilde{t_{r_j}}, t_{r_j}, p_{r_j})$ is released, denote the set of orders that are accepted by time $\widetilde{t_{r_j}}$ and are in conflict with r_j as

$$\Omega(r_j, B) = \{r \in B \mid |t_r - t_{r_j}| < 2t \text{ if } p_r = 1 - p_j, \text{ or } |t_r - t_{r_j}| < t \text{ if } p_r = p_{r_j}\}$$

BD Algorithm Description. At the high level, BD consists of two main procedures, the *order-accepting procedure* (see loop 1 of Algorithm 2) and the *car-dispatching procedure* (see loop 2 of Algorithm 2).

Algorithm 2. Balanced Dispatching (BD)

Input: k cars, online sequences of orders $R = \{r_1, r_2 ..., r_n\}$;
Output: a set of accepted orders;

1: **while** a new order $r_j = (\widetilde{t_{r_j}}, t_{r_j}, p_{r_j})$ is released **do**
2:    Find the index i such that $t_{r_j} \in T_i$, and let $j = p_{r_j}$;
3:    Set $\alpha = \lceil \frac{k}{2} \rceil$ if i is odd, $\alpha = \lfloor \frac{k}{2} \rfloor$ if i is even; set $\theta = \lceil \frac{\alpha}{2} \rceil$ if $j = 0$, and $\theta = \lfloor \frac{\alpha}{2} \rfloor$ if $j = 1$;
4:    **if** $|B_i^j| \leqslant \theta - 1$ and $|\Omega(r_j, B)| \leqslant k - 1$ **then**
5:        Accept r_j;
6:    **else**
7:        Reject r_j;
8:    **end if**
9: **end while**
10: **while** at time $t_{r_b} - t$ of $r_b \in B$ **do**
11:    **if** there are cars of those not in service in time interval $[t_{r_b}, t_{r_b} + t]$ that are located at p_{r_b} at t_{r_b} **then**
12:        Dispatch the last one of them that arrives at p_{r_b} no later than t_{r_b} to serve r_b;
13:    **else**
14:        Dispatch the last car that arrives at $1 - p_{r_b}$ no later than $t_{r_b} - t$ to serve r_b;
15:    **end if**
16: **end while**

– *In the order-accepting procedure.* BD always remains a budget $\alpha = \lceil \frac{k}{2} \rceil$ (resp. $\alpha = \lfloor \frac{k}{2} \rfloor$) of orders to accept in each odd-subscript (even-subscript) time slot. When a new order $r_j = (\widetilde{t_{r_j}}, t_{r_j}, p_{r_{new}})$ is released, BD goes to the order-accepting procedure to determine whether to accept r_j or not. Specifically, BD accepts r_j iff it meets the following two *acceptable conditions* together: (1) BD accepts at most $\theta = \lceil \frac{\alpha}{2} \rceil$ (resp. $\theta = \lfloor \frac{\alpha}{2} \rfloor$) orders that depart from location 0 (resp. location 1) in each time slot. (2) The number of accepted orders that are in conflict with r_j is at most k, i.e, $|\Omega(r_j, B)| \leqslant k - 1$.

- *Car-dispatching procedure.* A time t earlier than the start time t_{r_b} of each accepted order $r_b \in B$, BD goes to the car-dispatching procedure to dispatch a car to serve r_b by a greedy approach. Specifically, BD gives priority first to cars that can without-cost serve r_b and then to the car that is the last to become available to r_b.

Note that the number of orders in B that are in conflict is no more than k all the time, this implies the following Observations 1 since there are totally k cars in the system.

Observation 1. *Each accepted order by BD is able to be served by a car.*

Assume, *w.l.o.g.*, that there are totally n orders $R = \{r_1, r_2..., r_n\}$ that are released to the system within the first m time slots $T = \{T_1, T_2, ..., T_m\}$ considered in this paper. Further, denote R_i^0 and R_i^1 as the set of online orders that depart from location 0 and 1 in the time slot T_i. Accordingly, $R_i = R_i^0 \cup R_i^1$ denote the set of orders that are released in the time slot T_i, and further, $\bigcup_{i=1}^m R_i = R$. Denote O_i^0 (resp. O_i^1) as the set of accepted orders by OPT that depart from location 0 (resp. 1) within time slot T_i. Accordingly, denote $O_i = O_i^0 \cup O_i^1$ as the accepted orders by OPT that depart within time slot T_i, and further, $O = \bigcup_i O_i$ denote the set of orders that are accepted by OPT. Naturally, we have the following Observation 2. We also have the following Lemma 1 to be used in the bound analysis.

Observation 2. *In BD, the followings hold for each time slot $T_i \in T$, $|B_i^1| \geqslant \min\{\theta, |R_i^1|\} \geqslant \min\{\theta, |O_i^1|\}$, $|B_i^0| \geqslant \min\{\theta, |R_i^0|\} \geqslant \min\{\theta, |O_i^0|\}$, $\min\{k, |R_i^1|\} \geqslant |O_i^1|$, and $\min\{k, |R_i^0|\} \geqslant |O_i^0|$.*

Lemma 1. *For any four positive numbers N_1, N_2, M_1, M_2 with $M_1 \leqslant N_1, M_2 \leqslant N_2$, it holds that $\frac{N_1+N_2}{M_1+M_2} \leqslant \max\{\frac{N_1}{M_1}, \frac{N_2}{M_2}\}$.*

To present the upper bound on the competitive ratio of BD, we count the profit contributed by each served order in BD and OPT into the time slot from which the order departs, respectively. Recall that each without-cost (resp. with-cost) served order contributes a profit of γ (resp. $\gamma - \varsigma$) to the time slot from which the order departs. The profit of each time slot T_i achieved by BD and OPT, which are denoted by $I_{BD}(T_i)$ and $I_{OPT}(T_i)$, equals the sum of profit of served orders by BD and OPT that depart from the time slot, respectively.

Lemma 2. *For kS2L-V with $t \leqslant b_l$, the competitive ratio of BD is upper bounded by $\frac{k}{\lceil \frac{k}{4} \rceil}$ when there are only $\overrightarrow{10}$ orders released and $k \geqslant 4$.*

Lemma 3. *For kS2L-V with $t \leqslant b_l$, the competitive ratio of BD is upper bounded by $\frac{k}{\lceil \frac{k}{4} \rceil}$ when there are only $\overrightarrow{01}$ orders released to the system and $k \geqslant 4$.*

To further present the upper bound when both $\overrightarrow{01}$ and $\overrightarrow{10}$ orders are released to the system, we classify the time slots into two types, one type of time slots where there are $\overrightarrow{01}$ orders released, and the other type of time slots where there is no $\overrightarrow{01}$ order released. Denote $\widehat{T} = \{T_{\tau_1}, T_{\tau_2}, ..., T_{\tau_a}\}$ as the set of time slots in which there is at least one $\overrightarrow{01}$ order, where $\tau_1 < \tau_2 < \cdots < \tau_a$. Namely, there is no $\overrightarrow{01}$ order in each time slot of $T - \widehat{T}$. We observe the following.

Lemma 4. *For kS2L-V with $t \leqslant b_l$, the competitive ratio of algorithm BD is upper bounded by $\frac{2k}{\lfloor \frac{k}{4} \rfloor}$ when there are both $\overrightarrow{01}$ and $\overrightarrow{10}$ orders released to the system and $k \geqslant 4$.*

Proof. Now, we present the upper bound on the competitive ratio of BD by an induction method.

Case 1. Base Step. We first discuss the ratio of OPT's profit over BD's profit in the first τ_1 time slots $[0, \tau_1 \cdot t] = \bigcup_{i=1}^{\tau_1} T_i$. Since $R_1^0 = ... R_{\tau_1 - 1}^0 = \varnothing$, BD and OPT can only with-cost serve orders (if exist) that depart before T_{τ_1} respectively, because all the k cars are located at location 0 initially. In time slots $[0, (\tau_1 - 1) \cdot t] = \bigcup_{i=1}^{\tau_1 - 1} T_i$, we have

$$\frac{I_{OPT}(\bigcup_{i=1}^{\tau_1 - 1} T_i)}{I_{BD}(\bigcup_{i=1}^{\tau_1 - 1} T_i)} = \frac{(\gamma - \varsigma) \sum_{i=1}^{\tau_1 - 1} |O_i|}{(\gamma - \varsigma) \sum_{i=1}^{\tau_1 - 1} |B_i|} \tag{2}$$

At the start of T_{τ_1}, we note that BD has at least θ cars available at location 0 but no car available at location 1, because no $\overrightarrow{01}$ order is released before T_{τ_1} and BD always remain a budget θ of orders to accept in each time slot. This further implies that, in T_{τ_1}, BD can without-cost serve $\min\{\theta, |R_{\tau_1}^0|\}$ $\overrightarrow{01}$ orders and with-cost serve $\min\{\theta, |R_{\tau_1}^1|\}$ $\overrightarrow{10}$ orders (see Observation 2), implying

$$I_{BD}(T_{\tau_1}) = \gamma \cdot \min\{\theta, |R_{\tau_1}^0|\} + (\gamma - \varsigma) \cdot \min\{\theta, |R_{\tau_1}^1|\} \tag{3}$$

On the OPT's side, in T_{τ_1}, OPT without-cost serves $|O_{\tau_1}^0| \leqslant \min\{k, |R_{\tau_1}^0|\}$) $\overrightarrow{01}$ orders and with-cost serves $|O_{\tau_1}^1| \leqslant \min\{k - |O_{\tau_1}^0|, |R_{\tau_1}^1|\}$) $\overrightarrow{10}$ orders, implying

$$\begin{aligned}
I_{OPT}(T_{\tau_1}) &= \gamma \cdot |O_{\tau_1}^0| + (\gamma - \varsigma) \cdot |O_{\tau_1}^1| \\
&\leqslant \gamma \cdot \min\{k, |R_{\tau_1}^0|\} + (\gamma - \varsigma) \cdot \min\{k - \min\{k, |R_{\tau_1}^0|\}, |R_{\tau_1}^1|\} \\
&\leqslant \gamma \cdot \min\{k, |R_{\tau_1}^0|\} + (\gamma - \varsigma) \cdot \min\{\max\{k - |R_{\tau_1}^0|, 0\}, |R_{\tau_1}^1|\} \\
&\leqslant \gamma \cdot \min\{k, |R_{\tau_1}^0|\} + (\gamma - \varsigma) \cdot \min\{k, |R_{\tau_1}^1|\}
\end{aligned} \tag{4}$$

Hence, we have

$$\begin{aligned}
\rho_{BD}(\bigcup_{i=1}^{\tau_1} T_i) &= \frac{I_{OPT}(\bigcup_{i=1}^{\tau_1} T_i)}{I_{BD}(\bigcup_{i=1}^{\tau_1} T_i)} \\
&\leqslant \frac{(\gamma - \varsigma) \sum_{i=1}^{\tau_1 - 1} |O_i| + \gamma \cdot \min\{k, |R_{\tau_1}^0|\} + (\gamma - \varsigma) \cdot \min\{k, |R_{\tau_1}^1|\}}{(\gamma - \varsigma) \sum_{i=1}^{\tau_1 - 1} |B_i| + \gamma \cdot \min\{\theta, |R_{\tau_1}^0|\} + (\gamma - \varsigma) \cdot \min\{\theta, |R_{\tau_1}^1|\}} \\
&\leqslant \frac{(\gamma - \varsigma) \sum_{i=1}^{\tau_1 - 1} |O_i| + \gamma \cdot \min\{k, |R_{\tau_1}^0|\} + (\gamma - \varsigma) \cdot \min\{k, |R_{\tau_1}^1|\}}{(\gamma - \varsigma) \sum_{i=1}^{\tau_1 - 1} \{\theta, |O_i|\} + \gamma \cdot \min\{\theta, |R_{\tau_1}^0|\} + (\gamma - \varsigma) \cdot \min\{\theta, |R_{\tau_1}^1|\}} \\
&\leqslant \frac{k}{\theta} \leqslant \frac{k}{\lfloor \frac{k}{4} \rfloor}
\end{aligned} \tag{5}$$

in which the first inequality holds by (2)–(4), the second inequality holds by Observation 2 and the third inequality holds by $O_i \leqslant k$ for each $i \in \{1, \cdots, \tau_1\}$.

Case 2. Induction Step. Suppose the ratio holds for the first τ_j time slots $[0, \tau_j \cdot t] = \bigcup_{i=1}^{\tau_j} T_i$, we now prove that the ratio holds in the first τ_{j+1} time slots $[0, \tau_{j+1} \cdot t] = \bigcup_{i=1}^{\tau_{j+1}} T_i$ as well. We further discuss two cases.

Case 2.1. There are time slots between T_{τ_j} and $T_{\tau_{j+1}}$, say $\{T_{\tau_j+1}, \cdots, T_{\tau_{j+1}-1}\}$. Clearly, $R_{\tau_j+1}^0 = \cdots = R_{\tau_{j+1}-1}^0 = \varnothing$. In T_{τ_j+1}, BD accepts $\min\{\theta, |R_{\tau_j+1}^1|\}$ $\overrightarrow{10}$ orders by Observation 2, implying

$$I_{BD}(T_{\tau_j+1}) \geqslant (\gamma - \varsigma) \cdot \min\{\theta, |R_{\tau_j+1}^1|\} \tag{6}$$

In time slots $\bigcup_{i=\tau_j+2}^{\tau_{j+1}-1} T_i$, BD can accept $\sum_{i=\tau_j+2}^{\tau_{j+1}-1} \min\{\theta, |R_i^1|\}$ $\overrightarrow{10}$ orders, in which the first $\min\{\theta, |R_{\tau_j}^0|, \sum_{i=\tau_j+2}^{\tau_{j+1}-1} \min\{\theta, |R_i^1|\}\}$ orders (in non-decreasing order of departure time) are able to be without-cost served by the cars that are dispatched to $\overrightarrow{01}$ orders in $T_{\tau_j}^1$. Hence, BD earns a profit,

$$I_{BD}\left(\bigcup_{i=\tau_j+2}^{\tau_{j+1}-1} T_i\right)$$

$$\geqslant \gamma \cdot \min\{\theta, |R_{\tau_j}^0|, \sum_{i=\tau_j+2}^{\tau_{j+1}-1} \min\{\theta, |R_i^1|\}\}$$

$$+ (\gamma - \varsigma) \cdot \left(\sum_{i=\tau_j+2}^{\tau_{j+1}-1} \min\{\theta, |R_i^1|\} - \min\{\theta, |R_{\tau_j}^0|, \sum_{i=\tau_j+2}^{\tau_{j+1}-1} \min\{\theta, |R_i^1|\}\}\right) \tag{7}$$

$$= \varsigma \cdot \min\{\theta, |R_{\tau_j}^0|, \sum_{i=\tau_j+2}^{\tau_{j+1}-1} \min\{\theta, |R_i^1|\}\} + (\gamma - \varsigma) \sum_{i=\tau_j+2}^{\tau_{j+1}-1} \min\{\theta, |R_i^1|\}$$

As all the cars assigned to serve orders departing in the time slots $\bigcup_{i=\tau_j+1}^{\tau_2-1} T_i$ will reach location 0 after the service, at least θ cars reach location 0 either before or at the start time of $T_{\tau_{j+1}}$. Accordingly, BD can without-cost serve $\min\{\theta, |R_{\tau_{j+1}}^0|\}$ $\overrightarrow{01}$ orders that depart in $T_{\tau_{j+1}}$, implying

$$I_{BD}(T_{\tau_{j+1}}) \geqslant \gamma \cdot \min\{\theta, |R_{\tau_{j+1}}^0|\} + (\gamma - \varsigma) \cdot \min\{\theta, |R_{\tau_{j+1}}^1|\} \tag{8}$$

On the OPT's side, we note that only by those cars that are assigned with $\overrightarrow{01}$ orders in T_{τ_j} can the $\overrightarrow{10}$ orders departing within $\bigcup_{i=\tau_j+1}^{\tau_{j+1}-1} T_i$ be without-cost served. Hence, in time slots $\bigcup_{i=\tau_j+1}^{\tau_{j+1}-1} T_i$, OPT can without-cost serve at most

$$\min\{|O_{\tau_j}^0|, \sum_{i=\tau_j+1}^{\tau_{j+1}-1} |O_i^1|\} \leqslant \min\{k, |R_{\tau_j}^0|, \sum_{i=\tau_j+1}^{\tau_{j+1}-1} \min\{k, |R_i^1|\}\} \tag{9}$$

---

[1] We note that each with-cost served order occupies at most three consecutive time slots of its assigned car, including the time slot from which the order departs and its two neighbours.

$\overrightarrow{10}$ orders, implying

$$I_{OPT}\left(\bigcup_{i=\tau_j+1}^{\tau_{j+1}-1} T_i\right)$$

$$\leqslant \gamma \cdot \min\{k, R_{\tau_j}^0, \sum_{i=\tau_j+1}^{\tau_{j+1}-1} \min\{k, |R_i^1|\}\}$$

$$+ (\gamma - \varsigma) \cdot \left(\sum_{i=\tau_j+1}^{\tau_{j+1}-1} \min\{k, |R_i^1|\} - \min\{k, |R_{\tau_j}^0|, \sum_{i=\tau_j+1}^{\tau_{j+1}-1} \min\{k, |R_i^1|\}\}\right)$$

$$= \varsigma \cdot \min\{k, R_{\tau_j}^0, \sum_{i=\tau_j+1}^{\tau_{j+1}-1} \min\{k, |R_i^1|\}\} + (\gamma - \varsigma) \cdot \sum_{i=\tau_j+1}^{\tau_{j+1}-1} \min\{k, |R_i^1|\}$$

(10)

in which the inequality holds by (8). In $T_{\tau_{j+1}}$, we have

$$I_{OPT}(T_{\tau_{j+1}}) \leqslant \gamma \cdot (|O_{\tau_{j+1}}^1| + |O_{\tau_{j+1}}^0|) \leqslant \gamma \cdot k \qquad (11)$$

Substituting (6)–(11) in the left-hand term of the following (12), we have

$$\frac{I_{OPT}(\bigcup_{i=\tau_j+1}^{\tau_{j+1}-1} T_i) + I_{OPT}(T_{\tau_{j+1}})}{I_{BD}(\bigcup_{i=\tau_j+1}^{\tau_{j+1}-1} T_i) + I_{BD}(T_{\tau_{j+1}})} \leqslant \frac{k}{\theta} \qquad (12)$$

Therefore,

$$\rho_{BD} = \frac{I_{OPT}(\bigcup_{i=1}^{\tau_{j+1}} T_i)}{I_{BD}(\bigcup_{i=1}^{\tau_{j+1}} T_i)}$$

$$\leqslant \max\{\frac{k}{\theta}, \frac{I_{OPT}(\bigcup_{i=\tau_j+1}^{\tau_{j+1}-1} T_i) + I_{OPT}(T_{\tau_{j+1}})}{I_{BD}(\bigcup_{i=\tau_j+1}^{\tau_{j+1}-1} T_i) + I_{BD}(T_{\tau_{j+1}})}\}$$

$$\leqslant \frac{k}{\theta}$$

in which the first inequality holds by Lemma 1, and the second inequality holds by (12).

Case 2.2. There is no time slot between T_{τ_j} and $T_{\tau_{j+1}}$, i.e., $\tau_{j+1} = \tau_j + 1$. We note that BD accepts at least $\min\{\theta, R_i^0\}$ (resp. $\min\{\theta, R_i^1\}$) $\overrightarrow{01}$ (resp. $\overrightarrow{10}$) orders in each time slot T_i, implying that BD has at least $\min\{\theta, R_i^0\}$ (resp. $\min\{\theta, R_i^1\}$) cars that potentially without-cost serve $\overrightarrow{10}$ (resp. $\overrightarrow{01}$) orders that depart from time slot T_{i+1}. By a similar analysis as (5) in Case 1, we have

$$\rho_{BD}\left(\bigcup_{i=1}^{\tau_{j+1}} T_i\right) = \frac{I_{OPT}(\bigcup_{i=1}^{\tau_j-1} T_i) + I_{OPT}(T_{\tau_j}) + I_{OPT}(T_{\tau_{j+1}})}{I_{BD}(\bigcup_{i=1}^{\tau_j-1} T_i) + I_{BD}(T_{\tau_j}) + I_{BD}(T_{\tau_{j+1}})} \leqslant \frac{k}{\theta} \qquad (13)$$

By Cases 1-2, the competitive the ratio of BD is upper bounded by $\frac{k}{\lfloor\frac{k}{4}\rfloor}$ since

$$\theta \geqslant \left\lfloor\frac{\lfloor\frac{k}{2}\rfloor}{2}\right\rfloor \geqslant \lfloor\frac{k}{4}\rfloor.$$

By Lemma 2, Lemma 3 and Lemma 4, we have the following Theorem 9.

Theorem 9. *For kS2L-V with the setting $t \leqslant b_l$, the competitive ratio of algorithm BD is no more than $\frac{k}{\lfloor \frac{k}{4} \rfloor}$ if $k \geqslant 4$.*

4 Conclusions

This paper studies the online order-dispatching problem in the car-sharing system (CSS) in which there are only two locations (in which one is the departure location while the other is the destination). Compared to the previous literature, we consider general k cars and the empty-movement cost simultaneously in this paper. With regard to the booking time and the cost of empty movement, we studied different settings of the problem. Our contribution is two-fold. First, we show lower bounds on the competitive ratio for the considered settings, respectively. Then, we propose an online algorithm named GD that performs optimally, in terms of the competitive ratio, in most of our studied settings. For the setting $t \leqslant b_l$ in which GD cannot guarantee optimal performance, we propose another feasible online algorithm named BD with its competitive ratio proved to be close to the lower bound. In the future work, it would be interesting to extend the two-location-graph to a more general graph, such as a grid graph which usually models the traffic network in cities.

Acknowledgment. We thank Minming Li and Kelin Luo for many improvements in this work. We also thank anonymous reviewers for their helpful comments and suggestions.

References

1. Santi, P., Resta, G., Szell, M., Sobolevsky, S., Strogatz, S.H., Ratti, C.: Quantifying the benefits of vehicle pooling with shareability networks. Proc. Natl. Acad. Sci. **111**(37), 13290–13294 (2014)
2. Luo, K., Erlebach, T., Xu, Y.: Car-sharing between two locations: online scheduling with flexible advance bookings. In: Wang, L., Zhu, D. (eds.) COCOON 2018. LNCS, vol. 10976, pp. 242–254. Springer, Cham (2018). https://doi.org/10.1007/978-3-319-94776-1_21
3. Luo, K., Erlebach, T., Xu, Y.: Online scheduling of car-sharing requests between two locations with many cars and flexible advance bookings. In: 29th International Symposium on Algorithms and Computation. Schloss Dagstuhl-Leibniz-Zentrum fuer Informatik (2018)
4. Luo, K., Erlebach, T., Xu, Y.: Car-sharing between two locations: online scheduling with two servers. In: 43rd International Symposium on Mathematical Foundations of Computer Science, Liverpool, UK (2018)
5. Dickerson, J.P., Sankararaman, K.A., Srinivasan, A., Xu, P.: Allocation problems in ride-sharing platforms: online matching with offline reusable resources. In: 32nd AAAI Conference on Artificial Intelligence (2018)

6. Zhao, B., Xu, P., Shi, Y., Tong, Y., Zhou, Z., Zeng, Y.: Preference-aware task assignment in on-demand taxi dispatching: an online stable matching approach. In: 33th AAAI Conference on Artificial Intelligence (2019)
7. Bei, X., Zhang, S.: Algorithms for trip-vehicle assignment in ride-sharing. In: 32nd AAAI Conference on Artificial Intelligence (2018)
8. Huang, Z., Kang, N., Tang, Z.G., Wu, X., Zhang, Y., Zhu, X.: How to match when all vertices arrive online. In: the 50th Annual ACM SIGACT Symposium on Theory of Computing, pp. 17–29. ACM (2018)
9. Huang, Z.Y., Kang, N., Tang, G.T., Wu, X.W., Zhang, Y.H., Zhu, X.: Fully online matching. J. ACM **67**(3), Article no. 17 (2020). https://doi.org/10.1145/3390890
10. Borodin, A., El-Yaniv, R.: Online Computation and Competitive Analysis. Cambridge University Press, Cambridge (2005)

Buffer Minimization with Conflicts on a Line

Felix Höhne[✉] and Rob van Stee

Department of Mathematics, University of Siegen, Siegen, Germany
{felix.hoehne,rob.vanstee}@mathematik.uni-siegen.de

Abstract. In the buffer minimization in multiprocessor systems with conflicts or simply *buffer minimization* problem, a multi-processor system is modelled as an undirected graph. A conflict occurs if two processors are connected by an edge. Conflicting processors can not run at the same time. At any time, load may arrive on one or more processors. Incoming workload is stored in an input buffer and a machine that is running reduces its workload at a constant rate. The goal is to find a schedule that minimizes the maximum workload over all machines. We consider the special case where the graph is a path and give bounds on the competitive ratio for small graph sizes, including a tight bound of 9/4 for a path with 4 nodes. We give a general lower bound of 12/5. We also consider online algorithms that have resource augmentation on their speed, and give a $(1 + \varepsilon)$-speed $(1/\varepsilon + 3)$-competitive algorithm.

Keywords: Online algorithms · Scheduling · Conflict graph

1 Introduction

Scheduling is a fundamental problem in computer science. It played an important part in the development of approximation and online algorithms. One of the earliest online algorithms was designed for makespan scheduling in the 1960s [4]. Many objective functions for online scheduling have been studied over the years. In the well-known area of throughput scheduling, the goal is to maximize the weighted or unweighted number of completed jobs (the throughput). Each job has a deadline and we only count the jobs that complete by their deadlines [7].

In the current paper, we consider a variation of this problem in which jobs do not have deadlines. Instead, jobs that arrive are stored in a buffer of some size until they can be processed. The buffer minimization in multiprocessor systems with conflicts or simply *buffer minimization* problem was first introduced by Chrobak et al. [2]. A sequence of tasks needs to be scheduled in a multi-processor system with conflicts. A conflict occurs if two processors share a common resource that they cannot both access at the same time. The multi-processor system is modelled as an undirected graph where the processors are the nodes of the graph and are in conflict if they are connected by an edge. Without loss of generality, the conflict graph is connected.

© Springer Nature Switzerland AG 2020
M. Li (Ed.): FAW 2020, LNCS 12340, pp. 62–73, 2020.
https://doi.org/10.1007/978-3-030-59901-0_6

In this paper we consider the special case where this graph is a path. This means all processors lie next to each other and two adjacent processors cannot run at the same time. At any time tasks may arrive on a processor adding to this processor's workload. Each processor stores its workload in an input buffer. The goal is to find a scheduling strategy that minimizes the maximum buffer size of all processors.

An online algorithm processes the workload without knowledge of future tasks. We are interested in the competitive ratio, that is the ratio between the maximum buffer size of an online algorithm and the buffer size of an optimal offline algorithm in the worst case. Chrobak et al. [2] provide results for graphs, including K_n and trees as well as some competitive ratios for graphs with up to four vertices. In particular, they show that GREEDY is $\frac{5}{2}$-competitive on the graph that has four vertices on a line. One of the main results in this paper is providing the exact competitive ratio for this case.

1.1 Overview of Results

In this paper, we assume that the size of the offline buffer is known and we scale it to 1. Chrobak et al. [2] also consider the case where the offline buffer is not known and show that the competitive ratio in that case is at most 1 higher.

We provide a tight bound of $\frac{9}{4}$ on the competitive ratio for the path with four machines. For five machines, we give a $\frac{5}{2}$-competitive algorithm and a lower bound of $\frac{16}{7}$. For large m, we show that no algorithm can be better than $\frac{12}{5}$-competitive on m machines. We also give a $(\frac{1}{\varepsilon}+3)$-competitive algorithm, whose machines run at speed $1 + 2\varepsilon$. The following table sums up our and previous bounds on the competitive ratio for different path sizes.

Number of machines	2	3	4	5	m
Lower bound	3/2 [2]	2 [2]	9/4	16/7	12/5
Upper bound	3/2 [2]	2 [2]	9/4	5/2	$(m+1)/2$ [2]

Before this paper, the best known lower bound on general connected graphs (of any size) was only 2, whereas the best known upper bound is linear in the diameter of the graph (the length of the path, in our case).

1.2 Related Literature

The problem was first introduced by Chrobak et al. [2] They show that the competitive ratio is finite for all graphs. Furthermore they have results for the complete graph K_n and they show that the competitive ratio is exactly H_n, the nth harmonic number. For trees they achieve a competitive ratio of at most $\Delta/2 + 1$, where Δ is the tree diameter. They also give competitive ratios of GREEDY on small graphs including a competitive ratio of $\frac{5}{2}$ on the graph with 4 machines in a line.

A similar model is studied by Irani and Leung [5,6], who also introduce a conflict graph. However they study the conflict between jobs not processors and are minimizing the makespan. They show that even for paths no algorithm is better than $\Omega(n)$-competitive.

Bodlaender et al. [1] study a similar scenario in form of a two player game in which one player wants to keep a water-bucket system stable, whereas the other player wants to cause overflows. In their case the restriction is not that the player can only remove water from an independent set of buckets, instead the restriction is that the player can only empty subsets of consecutive buckets.

More recently, Gasieniec et al. [3] introduced the Bamboo Garden Trimming Problem, which is similiar to our problem, but does not feature any conflicts. A robot may trim only one bamboo at the end of each day, while the other bamboos grow at some rate. The goal is to find a schedule that maintains the elevation of the garden as low as possible.

2   Definitions

We are given m machines in a line numbered from 1 to m. We denote the workload of a machine by a_i, thus the state of an online algorithm can be described by the vector $a = (a_1, ..., a_m)$. We also call this a configuration. We denote the state or configuration of the optimal offline algorithm with the vector $z = (z_1, ..., z_m)$. Load vectors are never negative. We define the delay of a machine as $d_i := a_i - z_i$. Initially $a = z = 0$.

In a *discrete schedule*, timesteps are of size ε and during each ε-step the algorithm may choose an independent set I of machines and reduce the load of all machines (that have load) in this set by ε. Tasks arrive only before or after each timestep.

We also sometimes use a *continuous schedule*. Tasks may arrive at any time and machines reduce their workload continuously at a certain rate. This means two adjacent machines can share processing power both working at speed $\frac{1}{2}$ instead of just one of the machines working at any given time. The combined workrate of two adjacent machines can not exceed 1.

Both definitions are equivalent, meaning any discrete schedule is a special case of a continuous schedule and any continuous schedule can be approximated by a discrete schedule while increasing the buffer size by at most ε [2].

The algorithms in this paper consider groups of machines. A set S of adjacent machines is called a group if all machines in S have load and no machine adjacent to S has load.

3   The Competitive Ratio of the Path with Four Machines

In this section we determine the competitive ratio of the path with four machines. We begin by presenting a $\frac{9}{4}$-competitive algorithm:

Algorithm 1.

1. Run any group of size 1 at full speed.
2. If there is a group of size 2 in the middle, meaning $a_1 = a_4 = 0$, run the machine with higher load.
3. Else if there is a group of size 2 at the edge, run the more central machine, unless the other machine has load more than $\frac{5}{4}$. In that case run the other one.
4. If there is a group of size 3, run two machines unless the middle one has load more than $\frac{5}{4}$. In that case run the middle one.
5. If all machines have load, run the best parity (the parity that includes the machine with highest load), unless both outer machines have load more than $\frac{5}{4}$. In that case run the outer machines.

Lemma 1. *For Algorithm 1, the following invariants hold at all times.*

(I1) $d_1 + d_2 \leq \frac{3}{4}$, (I2) $d_2 + d_3 \leq \frac{1}{2}$, (I3) $d_3 + d_4 \leq \frac{3}{4}$, (I4) $d_1 + d_2 + d_3 + d_4 \leq 1$

Proof. Whenever all machines are empty, all invariants hold. When load arrives the invariants are obviously preserved. We consider task execution.

Observation 1. *If* ALG *(any algorithm) runs a machine at full speed, invariants (I1) to (I3) involving this machine are maintained.*

ALG reduces the load on that machine by ε during the ε-step, while OPT reduces the load of any two adjacent machines by at most ε, thus the invariant is maintained.

Observation 2. *Whenever any two machines are running, (I4) is maintained.*

ALG reduces the overall load by 2ε, while OPT can increase the overall delay by at most 2ε, therefore (I4) is maintained.

We show that when executing tasks, all invariants are maintained at all times. We consider all possible configurations of the load vector. In every situation some invariants are maintained because of the observations above. Given these invariants we then show that the remaining invariants also hold in that situation and are therefore also maintained. The complete proof can be found in the full version.   □

Theorem 1. *Algorithm 1 is $\frac{9}{4}$-competitive on the path with four machines.*

Proof. We first show that $d_1 < \frac{5}{4}$ holds at all times. Whenever $a_1 < \frac{5}{4}$, $d_1 < \frac{5}{4}$ holds. If $a_1 > \frac{5}{4}$ and machine 1 is running, d_1 does not increase. We only need to consider the case where $a_1 > \frac{5}{4}$ and machine 1 is not running. In this case $a_2 > \frac{5}{4}$ (Rules 4 and 5 of Algorithm 1). It follows that $d_2 > \frac{1}{4}$ and therefore $d_1 < \frac{1}{2}$ by (I1). This means $d_1 < \frac{5}{4}$ holds at all times.

We now show that $d_2 < \frac{5}{4}$. Again we only need to consider the case, where $a_2 > \frac{5}{4}$, but machine 2 is not running. In this case $a_1 > \frac{5}{4}$ or $a_3 > \frac{5}{4}$ (Rule 3 and 5 of Algorithm 1). It follows that $d_1 > \frac{1}{4}$ and therefore $d_2 < \frac{1}{2}$ by (I1) or $d_3 > \frac{1}{4}$ and therefore $d_2 < \frac{1}{4}$ by (I2). This means $d_2 < \frac{5}{4}$ holds at all times.

By symmetry the delay on each machine i is at most $d_i = a_i - z_i \leq \frac{5}{4}$. Adding $z_i - 1 \leq 0$ yields $a_i \leq \frac{9}{4}$. Therefore the algorithm is $\frac{9}{4}$-competitive on the path with four machines. $\square$

We now present a matching lower bound of $\frac{9}{4}$ on the competitive ratio of the path with four machines.

Lemma 2. *Let the competitive ratio of* ALG *be equal to* $2 + x$. *Whenever* $z_i = 0$ *is possible we have* $a_i \leq 1 + x$.

Proof. Otherwise one unit of load arriving on machine i would yield a ratio higher than $2 + x$. $\square$

Lemma 3. *If* ALG *has a delay of* x *on two adjacent machines, we can reach a state where* ALG *has load at least* x *on those machines and all machines of* OPT *are empty.*

Proof. Let the two adjacent machines be i and $i + 1$. No load arrives until OPT has removed its load on machines i and $i + 1$. Since there was a delay of x on those machines, there is still at least x load left once OPT's machines i and $i + 1$ are empty. Then load arrives on machine i and OPT runs all machines that have the same parity as i. After one timestep, this parity is empty for OPT, as it has load at most 1 on each machine at all times, and the total load of ALG on i and $i + 1$ has not changed. Repeat this for the other parity. $\square$

Theorem 2. *No algorithm can be better than* $\frac{9}{4}$-*competitive on the path with four machines.*

Proof. Let the competitive ratio of ALG be equal to $2 + x$. Suppose that $x < \frac{1}{4}$. We start with both OPT's and ALG's machines empty. We present two phases, consisting of repeating sets of similar inputs. The first phase creates an initial delay on the middle two machines. Based on that, the second phase forces a competitive ratio that is unbounded for $x < \frac{1}{4}$, which contradicts the assumption.

Phase 1 (see Table 1): Repeating the following set of inputs forces a total load of (almost) $1 - 2x$ on the middle two machines of ALG in the limit, with OPT's machines being empty.

Let $L = a_2 + a_3$ be the combined load of the middle two machines before the first time step. Load arrives after every timestep i. We denote the time right before load arrives with i^- and the time right after with i.

Time 0: At time 0, one unit of load arrive on machines 2 and 3 each. Thus a_2 and a_3 increase by 1 each, while during the first timestep $a_2 + a_3$ decreases by at most 1. This means at time 1^-, $a_2 \geq \frac{L+1}{2}$ or $a_3 \geq \frac{L+1}{2}$. W.l.o.g., let $a_2 \geq \frac{L+1}{2}$.

Table 1. Phase 1: The notation i^- refers to the state at time i just before new jobs arrive, i refers to the state after the job arrivals. The boxed entries are lower bounds for two adjacent machines. Those marked with $\leq$ are upper bounds.

Time\Machines	Loads of ALG				Loads of OPT				Next input
	1	2	3	4	1	2	3	4	
0^-	...	$\boxed{L}$		...	0	0	0	0	0110
0	...	$\boxed{L+2}$		...	0	1	1	0	
1^-	0	$\boxed{L+1}$		0	0	0	1	0	1000
1	1	$\boxed{L+1}$		0	1	0	1	0	
2^-	$\boxed{\frac{L+1}{2}}$		...	0	0	0	0	0	1100
2	$\boxed{\frac{L+5}{2}}$		...	0	1	1	0	0	
3^-	$\leq 1+x$	$\boxed{\frac{L+1}{2}-x}$		0	1	0	0	0	0010
3	$\leq 1+x$	$\boxed{\frac{L+3}{2}-x}$		0	1	0	1	0	
4^-	x	$\boxed{\frac{L+1}{2}-x}$		0	0	0	0	0	0110

During this timestep we allow ALG to process all existing loads on machines 1 and 4 from previous iterations (which only makes it easier for ALG).

Time 1: One unit of load arrives on machine 1. At time 2^-, $a_1 + a_2 \geq \frac{L+1}{2}$ and OPT has configuration $(0, 0, 0, 0)$.

Time 2: One unit of load arrives on machines 1 and 2 each, while $a_1 + a_2$ decreases by at most 1 during the following timestep. This means $a_1 + a_2 \geq 1 + \frac{L+1}{2}$ at time 3^-.

Since a possible configuration of OPT at this time is $(0, 1, 0, 0)$, $a_1 \leq 1 + x$ by Lemma 2. This implies $a_2 \geq \frac{L+1}{2} - x$ at time 3^-.

Time 3: One unit of load arrives on machine 3. Then at time 4^-, $a_2 + a_3 \geq \frac{L+1}{2} - x$ while the machines of OPT are again empty. We can now repeat the input if desired.

Repeating these four timesteps, we get a sequence $(L_n)_n$ with $L_{n+1} \geq \frac{L_n+1}{2} - x$.

As long as $L_n \leq 1 - 2x$, we have

$$1 - 2x - L_{n+1} \leq \frac{1}{2}(1 - 2x - L_n)$$

Since $L_0 = 0 \leq 1 - 2x$ we find $1 - 2x - L_{n+1} \leq \frac{1}{2^n}(1 - 2x)$, so $L_{n+1} \geq 1 - 2x - \frac{1}{2^n}(1 - 2x)$.

Phase 1 continues until $L_n > 6x - 1$. This is possible because $x < \frac{1}{4}$ and therefore $1 - 2x > 6x - 1$.

Phase 2 (see Table 2): Again let L be the load on the middle two machines. Let B be the load on all machines. We reset time to 0. This means after phase 1 we start phase 2 with a load of $L_0 > 6x - 1$.

Time 0: At time 0, one unit of load arrives on machines 2 and 3 each. During this timestep ALG can reduce $a_2 + a_3$ by 1. We again allow ALG to process all existing loads on machines 1 and 4 from previous iterations. For the total load at time 1^- (before new jobs arrive) we have

$$B \geq L + 1$$

with one of the machines (w.l.o.g. machine 2) having at least load $\frac{L+1}{2}$.

Table 2. Phase 2: The entries for machines 2 and 3 of ALG and the total load are lower bounds. Those marked with $\leq$ are upper bounds.

Time \Machines	Loads of ALG 1	2 and 3	4	Loads of OPT 1	2	3	4	Total load	Next input
0^-	...	L	...	0	0	0	0	L	0110
0	...	$L+2$	...	0	1	1	0	$L+2$	
1^-	0	$L+1$	0	0	0	1	0	$L+1$	1000
1	1	$L+1$	0	1	0	1	0	$L+2$	
2^-				0	0	0	0		1110
2				1	1	1	0		
3^-				0	1	0	0	$\frac{3}{2}L - x + \frac{3}{2}$	1011
3				1	1	1	1	$\frac{3}{2}L - x + \frac{3}{2}$	
4^-				$\frac{1}{2}$	$\frac{1}{2}$	$\frac{1}{2}$	$\frac{1}{2}$	$\frac{3}{2}L - x + \frac{5}{2}$	
4									1 and 4 continuous
5^-	$\leq 1 + x$	$\frac{3}{2}L - 3x + \frac{1}{2} \leq 1 + x$		1	0	0	1	$\frac{3}{2}L - x + \frac{5}{2}$	0110
...	...	$\frac{3}{2}L - 3x + \frac{1}{2}$	...	0	0	0	0		0110

Time 1 and 2: At time 1, one unit of load arrives on machine 1. After this step OPT's machines are empty. Then load arrives on machines 1, 2 and 3. At time 3^-, OPT has configuration $(0, 1, 0, 0)$, however $(1, 0, 1, 0)$ is also a possible configuration. Because $(1, 0, 1, 0)$ is a possibility, $a_2 \leq 1 + x$ at time 3 by Lemma 2. During these two timesteps, ALG therefore has to run machine 2 (alone, since machine 4 is empty) for a duration of at least $\frac{L+1}{2} + 1 - (1 + x) = \frac{L+1}{2} - x$. ALG can then run machines 1 and 3 for a duration of at most $2 - (\frac{L+1}{2} - x) = -\frac{1}{2}L + x + \frac{3}{2}$. This decreases the overall load by at most twice that amount. (It is less if machine 3 becomes empty.) For the total load at time 3^- we have $B \geq L + 5 - (\frac{L+1}{2} - x) - 2 \cdot (-\frac{1}{2}L + x + \frac{3}{2}) = \frac{3}{2}L - x + \frac{3}{2}$.

Time 3: At time 3, one unit of load arrives on machines 1, 3 and 4 each. This means all machines of OPT have load 1. OPT then uses processor sharing to equally decrease the load on all machines, ending in state $(\frac{1}{2}, \frac{1}{2}, \frac{1}{2}, \frac{1}{2})$ at time 4^-. However it is also possible for OPT to be in state $(0,1,1,0)$ at time 4^-. At time 3, three units of load arrive, but ALG can only remove two in one time unit. This means

$$B \geq \frac{3}{2}L - x + \frac{5}{2}. \tag{1}$$

at time 4^-.

Time 4: Then, *during* timestep 5, one unit of load arrives on machines 1 and 4. This happens continuously at a constant speed of 1 over the whole timestep.

OPT continues to use processor sharing, ending in the state $(1,0,0,1)$. However ALG needs to take into account that if OPT was in state $(0,1,1,0)$ at time 4^-, OPT could have continously run machines 1 and 4 during timestep 5. In this case it would still be in state $(0,1,1,0)$ at time 5^-.

The overall load added during this timestep is 2. This means at time 5^- we still have (1). Because the state $(0,1,1,0)$ is possible for OPT, $a_1 \leq 1 + x$ and $a_4 \leq 1 + x$ by Lemma 2. The rest of the load is on the middle two machines. Given our bound on B, the middle two machines have a load of at least

$$B - 2 \cdot (1 + x) \geq \frac{3}{2}L - 3x + \frac{1}{2}. \tag{2}$$

By Lemma 3 we can now reach a state where the middle two machines have load at least $\frac{3}{2}L - 3x + \frac{1}{2}$ and all of OPT's machines are empty.

We can now repeat this phase if desired.

Let $\varepsilon_n = L_n - (6x - 1)$. We show that $\varepsilon_{n+1} \geq \frac{3}{2}\varepsilon_n$. Since $L_{n+1} \geq \frac{3}{2}L_n - 3x + \frac{1}{2}$, we have

$$\varepsilon_{n+1} - \varepsilon_n = L_{n+1} - L_n = \frac{1}{2}L_n - 3x + \frac{1}{2} = \frac{1}{2}(6x - 1 + \varepsilon_n) - 3x + \frac{1}{2} = \frac{1}{2}\varepsilon_n,$$

hence $\varepsilon_{n+1} \geq \frac{3}{2}\varepsilon_n$. It follows that $L_n \geq L_0 + (\frac{3}{2})^n \varepsilon_0$. Since $\varepsilon_0 > 0$, this implies $(L_n)_n$ grows without bound, contradicting the assumption that the competitive ratio is $2 + x$.   □

4   Bounds on the Competitive Ratio for Five Machines

In this section we first present an algorithm that achieves a competitive ratio of $\frac{5}{2}$ and follow up with a lower bound of $\frac{16}{7}$.

Algorithm 2.

---

1. Run any group of size 1 at full speed.
2. If there is a group of size 2 at the edge run the more central machine, unless the other machine has load more than $\frac{3}{2}$.
3. Otherwise if there is a group of size 2 run the more central machine, unless the other machine has load more than 1.
4. If there is a group of size 3 run two machines unless the middle one has load more than $\frac{3}{2}$.
5. We consider a group of size 4, e.g. machines 1 to 4. We run machines 2 and 4 unless $a_1 > \frac{3}{2}$ or $a_3 > \frac{3}{2}$. In that case we run machines 1 and 3. Additionally if $a_1 > \frac{3}{2}$ and $a_4 > \frac{3}{2}$ we run machines 1 and 4. Machines 2 to 5 are symmetrical to machines 1 to 4.
6. If all machines have load we run the odd parity, unless $a_2 > \frac{3}{2}$ or $a_4 > \frac{3}{2}$. If $a_2 > \frac{3}{2}$ we run the even parity as long as $a_5 \leq \frac{3}{2}$. If $a_2 > \frac{3}{2}$ and $a_5 > \frac{3}{2}$ we run machines 2 and 5. The case of $a_4 > \frac{3}{2}$ is symmetrical to $a_2 > \frac{3}{2}$.

Lemma 4. *Algorithm 2 maintains the following invariants at all times.*

1. $d_i \leq \frac{3}{2}$
2. $d_i + d_{i+1} \leq 1$ *for* $i = 1, 4$
3. $d_i + d_{i+1} \leq \frac{1}{2}$ *for* $i = 2, 3$
4. $d_i + d_{i+1} + d_{i+2} + d_{i+3} \leq 1$

Proof. The complete proof can be found in the full version.   □

Theorem 3. *Algorithm 2 is $\frac{5}{2}$-competitive on P_5.*

Proof. Follows directly from invariant 1 in the lemma above.   □

We now present the lower bound. The construction for four machines forces a load of $1 - 2x$ on the middle two machines in phase 1. For five machines we can force a load of $1 - x$ on machine 3. Applying phase 2 of the construction for four machines with a higher starting value yields a higher lower bound.

Lemma 5. *For any competitive algorithm and any $\varepsilon \in [0, 1]$, there exists a set of inputs on P_5 and $n \in \mathbb{N}$, so that at time n, $a_2 \geq 1 - \varepsilon$ or $a_4 \geq 1 - \varepsilon$ and OPT is in configuration $(0, 0, 1, 0, 0)$.*

Proof. We present a repeating set of inputs and show that, if the algorithm never allows a load of at least $1 - \varepsilon$ on machines 2 or 4, the load on machine 3 grows without bound. The construction is essentially one from Chrobak et al. (Theorem 5.4 in [2]) Load arrives after every timestep i. We denote the time right before load arrives with i^- and the time right after with i.

Let $a_2 + a_3 = L$ and OPT's machines be empty at time t^-.

Time t: One unit of load arrives on machines 2 and 3 each, therefore $a_2 + a_3 = L + 2$. At time $(t+1)^-$, we have $a_2 + a_3 = L + 1$, with $a_3 \geq L + \varepsilon$, since $a_2 \leq 1 - \varepsilon$. OPT can be in the configuration $(0, 1, 0, 0, 0)$.

Time t + 1: One unit of load arrives on machine 4. At time $(t + 2)^-$, we have $a_3 + a_4 = L + \varepsilon$ and OPT's machines are empty.

Because of symmetry, we can now repeat these inputs starting from an initial load of $L' = L + \varepsilon$ on machines 3 and 4. After every repetition the initial load increases by ε and therefore grows without bound. Because the load on machines 2 and 4 is bounded by $1 - \varepsilon$ this means the load on machine 3 grows without bound. Therefore the algorithm is not competitive if it never allows a load of at least $1 - \varepsilon$ on machines 2 and 4.

This means any competitive algorithm has to allow a load of at least $1 - \varepsilon$ on machines 2 or 4 at some point. W.l.o.g. this happens at machine 2 at time t^-. OPT can then be in configuration $(0, 0, 1, 0, 0)$ instead of $(0, 1, 0, 0, 0)$. This concludes the proof.   □

Theorem 4. *No Algorithm can be better than $\frac{16}{7}$-competitive on P_5.*

Proof. Let the competitive ratio of ALG be equal to $2 + x$. Suppose that $x < \frac{2}{7}$.

W.l.o.g. let $a_2 \geq 1 - \varepsilon$ with OPT being in the state $(0, 0, 1, 0, 0)$. We can reach this state for any competitive algorithm because of Lemma 5. Let the time be 0^-.

The following set of inputs forces a combined load of at least $1 - x - \varepsilon$ on machines 2 and 3, while all machines of OPT are empty.

Time 0: One unit of load arrives on machine 1. At time 1^-, $a_1 + a_2 \geq 1 - \varepsilon$. OPT'S machines are empty.

Time 1: One unit of load arrives on machine 1 and 2 each. At time 2^-, $a_1 + a_2 \geq 2 - \varepsilon$. OPT is in configuration $(1, 0, 0, 0, 0)$. Additionally $a_1 \leq 1 + x$ because of Observation 2. This means $a_2 \geq 1 - x - \varepsilon$

Time 2: One unit of load arrives on machine 3. At time 3^-, $a_2 + a_3 \geq 1 - x - \varepsilon$. All machines of OPT are empty.

We use the achieved configuration as a starting configuration and run phase 2 (from the proof of the lower bound on P_4) on machines 1 to 4. During phase 2 the load on machines 2 and 3 grows without bound, as long as the initial load is more than $6x - 1$ and the side loads remain below $1 + x$. This is the case, because the initial load is arbitrarily close to $1 - x$ and $x < \frac{2}{7}$. This means no algorithm can achieve a competitive ratio better than $\frac{16}{7}$.   □

5   Results on m Machines

In this section we provide some results for m machines. The first result is an $(\frac{1}{\varepsilon} + 3)$-competitive algorithm, with machines running at speed $1 + 2\varepsilon$.

Lemma 6. *In any interval of length T, at most $T + 2$ load arrives on any two adjacent machines and at most $T + 1$ load arrives on any single machine.*

Proof. Even if the buffers of OPT are empty at the start of this interval, it can process only T load during the interval, therefore more than $T + 2$ load would make OPT's buffer overflow. Analogously, at most $T + 1$ load arrives on any single machine.   □

Algorithm 3.

Use phases of length $\frac{1}{\varepsilon}$.

- In phase 0 do nothing.
- In phases > 0 let L_i be the load that arrived on each *odd* machine i. Process this load during the first $\frac{L_i}{1+2\varepsilon}$ time in the next phase.
- Run each *even* machine whenever it can be run. This means it gets run at the end of each phase, after both of its neighbours have stopped processing.

Theorem 5. *Algorithm 3, whose machines run at $(1 + 2\varepsilon)$-speed, is $(\frac{1}{\varepsilon} + 3)$-competitive.*

Proof. W.l.o.g. consider machines 1 to 3 and let the current phase be j. In total the algorithm can process $\frac{1+2\varepsilon}{\varepsilon} = \frac{1}{\varepsilon} + 2$ load on machines 1 and 2, which means it can process everything that could possibly arrive during phase $j-1$ by Lemma 6.

The load on odd machines is highest at the start of the phase and is at most $\frac{1}{\varepsilon} + 1$, because no more load can arrive during phase $j - 1$ and all the load that arrived before phase $j - 1$ gets processed before phase j.

For the load on even machines let L_i be the load that arrived during the previous phase on machine i and consider machine 2. W.l.o.g. $L_1 > L_3$. At the end of a phase, its load is at most $\frac{1}{\varepsilon} + 2 - L_1$. By the time it starts processing machine 1 has run for $\frac{L_1}{1+2\varepsilon}$, which means $\frac{L_1}{1+2\varepsilon} + 1$ additional load can arrive on machine 2. Thus when machine 2 starts processing the load is at most $\frac{1}{\varepsilon} + 2 - L_1 + \frac{L_1}{1+2\varepsilon} + 1 = \frac{1}{\varepsilon} + 3 - \frac{2\varepsilon L_1}{1+2\varepsilon} \leq \frac{1}{\varepsilon} + 3$. $\qquad\square$

Finally, we present a lower bound of $\frac{12}{5}$ on m machines.

Theorem 6. *No algorithm can be better than $\frac{12}{5}$-competitive on m machines for m large enough.*

Proof. Let $x < \frac{2}{5}$.

We consider 4 adjacent machines numbered 1 to 4. We start with a load of L on machines 2 and 3, while all machines of OPT are empty. There is a set of inputs that forces a total load of at least $\frac{5}{2} - x + \frac{3}{2}L_n > \frac{21}{10} + \frac{3}{2}L_n$ on all machines combined, while OPT can be in state $(1, 0, 0, 1)$ as well as $(0, 1, 1, 0)$ (phase 2 of the proof of the lowerbound on P_4). If we reach a state where ALG has a delay of L_{n+1} on two neighbouring machines, we can reach a state where ALG has a load of L_{n+1} on these machines and all machines of OPT are empty by Lemma 3.

This means if, after the set of inputs, there is a delay of $L_{n+1} > L_n$ on either machine 1, machines 2 and 3, or machine 4 we can reach a state with a higher starting load L on two adjacent machines and can repeat the set of inputs. This is clearly true if this load is on machines 2 and 3. If the load is on machine 1 or machine 4, then we shift the set of considered machines to the left or right. This is possible, because we assume the initial set of four machines to be in the middle of the set of m machines. This means for m large enough, there can always be empty machines to the left or right of the considered set of four machines. Thus if there is a delay of $d_1 = L_{n+1}$ on machine 1 and OPT is in state $(0, 1, 1, 0)$, we remove machines 3 and 4 from the considered machines and add 2 machines to the left of machine 1. Then we update the numbering accordingly. We can then reach a state where the middle two machines have load L_{n+1} and all machines of OPT are empty.

Next, we choose the highest load out of a_1, $a_2 + a_3$ and a_4 as our next load L_{n+1}. Since the total load is at least $\frac{21}{10} + \frac{3}{2}L_n$ we have $L_{n+1} \geq (\frac{21}{10} + \frac{3}{2}L_n)/3 = \frac{7}{10} + \frac{1}{2}L_n$. Starting from load $L_0 = 0$ this recurrence can be expressed as $L_n = \frac{7}{5} - (\frac{1}{2})^n \frac{7}{5}$. Thus for large enough n (which may require many iterations and shifts and thus a large m) we get a load that is arbitrarily close to $\frac{7}{5}$ on the middle pair of a set of four machine, while all machines of OPT are empty. We can now apply phase 2 of the lower bound for four machines with a starting load of $\frac{7}{5}$. Since $x < \frac{2}{5}$, we have $\frac{7}{5} > 6x - 1$ and the load on the middle two machines grows without bound as long as the side loads remain below $1 + x$. This concludes the proof. $\Box$

References

1. Bodlaender, M.H.L., Hurkens, C.A.J., Kusters, V.J.J., Staals, F., Woeginger, G.J., Zantema, H.: Cinderella versus the wicked stepmother. In: Baeten, J.C.M., Ball, T., de Boer, F.S. (eds.) TCS 2012. LNCS, vol. 7604, pp. 57–71. Springer, Heidelberg (2012). https://doi.org/10.1007/978-3-642-33475-7_5
2. Chrobak, M., Csirik, J., Imreh, C., Noga, J., Sgall, J., Woeginger, G.J.: The buffer minimization problem for multiprocessor scheduling with conflicts. In: Orejas, F., Spirakis, P.G., van Leeuwen, J. (eds.) ICALP 2001. LNCS, vol. 2076, pp. 862–874. Springer, Heidelberg (2001). https://doi.org/10.1007/3-540-48224-5_70
3. Gąsieniec, L., Klasing, R., Levcopoulos, C., Lingas, A., Min, J., Radzik, T.: Bamboo garden trimming problem (perpetual maintenance of machines with different attendance urgency factors). In: Steffen, B., Baier, C., van den Brand, M., Eder, J., Hinchey, M., Margaria, T. (eds.) SOFSEM 2017. LNCS, vol. 10139, pp. 229–240. Springer, Cham (2017). https://doi.org/10.1007/978-3-319-51963-0_18
4. Graham, R.L.: Bounds on multiprocessing timing anomalies. SIAM J. Appl. Math. **17**(2), 416–429 (1969)
5. Irani, S., Leung, V.J.: Scheduling with conflicts, and applications to traffic signal control. In: Tardos, É. (ed.) Proceedings of the Seventh Annual ACM-SIAM Symposium on Discrete Algorithms, Atlanta, Georgia, USA, 28–30 January 1996, pp. 85–94. ACM/SIAM (1996)
6. Irani, S., Leung, V.J.: Probabilistic analysis for scheduling with conflicts. In: Saks, M.E. (ed.) Proceedings of the Eighth Annual ACM-SIAM Symposium on Discrete Algorithms, New Orleans, Louisiana, USA, 5–7 January 1997, pp. 286–295. ACM/SIAM (1997)
7. Sgall, J.: Open problems in throughput scheduling. In: Epstein, L., Ferragina, P. (eds.) ESA 2012. LNCS, vol. 7501, pp. 2–11. Springer, Heidelberg (2012). https://doi.org/10.1007/978-3-642-33090-2_2

Single Machine Scheduling Problem with a Flexible Maintenance Revisited

Dehua Xu[1(✉)], Limin Xu[1], and Zhijun Xu[2]

[1] School of International Economics and Business, Nanjing University of Finance
& Economics, Nanjing 210023, People's Republic of China
{xudehua,xulimin}@nufe.edu.cn
[2] Fuzhou Teachers' College, East China University of Technology, Fuzhou 344000,
People's Republic of China
zhjxu@ecut.edu.cn

Abstract. The main aim of this paper is to determine the worst-case
bound of an algorithm proposed for a single machine scheduling problem with a flexible maintenance, which is open for over seventeen years.
Sensitivity analysis on the initial start time of the maintenance activity
reveals an interesting phenomenon of bound jump, which shows that the
proposed algorithm is the best among a series of algorithms from the
worst-case bound point of view.

Keywords: Scheduling · Flexible maintenance · Worst-case analysis

1 Introduction

Machine scheduling plays an important role in manufacturing systems [2,4,23].
In order to maintain a good condition, it is usually necessary to shutdown a
machine for a period of time within a planning horizon due to regular inspection,
refueling, tool change and so on.

What follows is a single machine scheduling problem that was considered by
Yang et al. [32] seventeen years ago. There are n non-preemptive jobs J_1, J_2, ...,
J_n to be processed on a single machine which is subject to a flexible maintenance
activity with a fixed duration of r. The start time of the maintenance activity
is a decision variable that has to be determined in a prefixed interval $[s, t]$ with
$r \leq t - s$, where time s (t) is the earliest (latest) time at which the machine
starts (stops) its maintenance. The processing time of job J_i is p_i. All jobs are
available at time zero. The machine can process at most one job at a time and
can not process any jobs when it is undergoing maintenance. The objective is to
determine the start time of the maintenance activity and schedule all the jobs
to the machine such that the makespan, i.e., the completion time of the last
finished job, is minimized.

Supported by the Qinglan Project of Universities in Jiangsu and the Science and Technology Project of Jiangxi Education Department (GJJ170446).

M. Li (Ed.): FAW 2020, LNCS 12340, pp. 74–82, 2020.
https://doi.org/10.1007/978-3-030-59901-0_7

Yang et al. [32] first show that this problem is NP-hard and then propose a heuristic algorithm based on the well-known *longest processing time first* (LPT) rule [23]. They point out that the performance of the algorithm is quite satisfactory by computational experiments.

It is easy to see that this problem falls into the category of *scheduling with machine non-availability*, which is one of the hottest topics in the scheduling community. For related research on scheduling with a single maintenance activity, including some variants of the problem under consideration, interested readers are referred to Breit [5], Cheng and Wang [7], Detti et al. [8], Hnaien et al. [9], Kacem and Kellerer [11], Kacem and Levner [12], Lee [14], Low et al. [16], Luo and Chen [17], Luo et al. [18], Luo and Ji [19], Luo and Liu [20], Mosheiov and Sarig [21], Xu et al. [27]. For scheduling with multiple maintenance activities, interested readers are referred to Abdellafou et al. [1], Beaton et al. [3], Chen [6], Huo and Zhao [10], Li et al. [15], Perez-Gonzalez and Framinan [22], Wang et al. [24], Xu et al. [25,26,31], Xu and Xu [28,29], Xu and Yang [30], to name just a few.

The introduction of machine maintenance in no doubt makes it harder to make appropriate decisions in production management. As is well-known, worst-case bound (also known as worst-case ratio) is one of the main performance measures for approximation algorithms. It guarantees the maximum deviation of the results generated by an algorithm from the corresponding optimal ones. One may have noticed that there exists a *fully polynomial time approximation scheme* (FPTAS) for the scheduling problem under consideration based on the fact that there exists an FPTAS for the classical knapsack problem [13] by setting the capacity of the knapsack to $t - r$ while viewing each job as an item (item j with a weight of p_j and a profit of p_j). The desired schedule can be constructed by assigning the jobs that correspond to the items packed into the knapsack before the maintenance activity and the others after the maintenance activity while prohibiting any unforced idleness. The correctness can be checked by considering the relationships among the optimal total profit of the knapsack problem, the total profit generated by the FPTAS for the knapsack problem, and the earliest start time s of the maintenance activity case by case. For an NP-hard problem under consideration, the existence of an FPTAS is evidently important in theory while heuristics and meta-heuristics may work pretty well in practice. However, this does not mean that worst-case bound analysis of a particular algorithm is dispensable, especially for such a straightforward algorithm based on the well-known LPT rule for such a typical and fundamental scheduling problem studied in Yang et al. [32], which is open for over seventeen years. The worst-case bound analysis, especially the determination of the exact bound, of such an algorithm may enable one to understand the algorithm and the corresponding problem better, which is in turn beneficial to the design and analysis of some other more sophisticated algorithms.

The main aim of this paper is to determine the exact worst-case bound of the algorithm proposed by Yang et al. [32] for the scheduling problem under consideration. We show that the worst-case bound of the algorithm for the scheduling problem under consideration is 4/3 and it is tight. Moreover, we show that the exact worst-case bound of any other algorithm which is the same as the algorithm proposed by Yang et al. [32] but with a different initial start time of the maintenance activity is always equal to 2, which implies that the algorithm proposed by Yang et al. [32] is the best among this series of algorithms from the worst-case bound point of view.

2   The Exact Worst-Case Bound

Now, let $v = t - r$. Clearly, v is the latest time at which the machine starts its maintenance activity. As stated earlier, Yang et al. [32] propose a heuristic algorithm for the single scheduling problem under consideration. For the convenience of description, we denote their algorithm by LPT_v. The main idea of their algorithm is to try jobs one by one according to decreasing order of their processing times and assign as many jobs as possible before time v. Specifically, their algorithm can be restated as follows.

Algorithm LPT$_v$

Step 1. Re-index all the jobs in decreasing order of their processing times, i.e., $p_1 \geq p_2 \geq \ldots \geq p_n$. Let the maintenance activity start at time v. Let $i = 1$ and $\delta = v$.

Step 2. If job $p_i \leq \delta$, then schedule job J_i before the maintenance activity as early as possible without changing the schedule of the other scheduled jobs (if any) and let $\delta = \delta - p_i$, otherwise schedule job J_i after the maintenance activity as early as possible without changing the schedule of the other scheduled jobs (if any). Let $i = i + 1$.

Step 3. If $i \leq n$, then go to Step 2.

Step 4. If $\delta > 0$, then shift the maintenance activity and each of the jobs scheduled after the maintenance activity forward (to the direction of the origin, i.e., time zero) by $\min\{\delta, v - s\}$. Output the current schedule as the final schedule.

Let $\Lambda = \sum_{i=1}^{n} p_i$ and $p_{\min} = \min_{1 \leq i \leq n}\{p_i\}$. Clearly, if $\Lambda \leq v$, then the scheduling problem under consideration becomes trivial since an optimal schedule can be obtained by processing the jobs one by one in any order consecutively from time zero [32]. Another trivial case is the case where $p_{\min} > v$ for which an optimal schedule can be obtained by letting the maintenance activity start at time s and processing the jobs one by one in any order consecutively from time $s + r$ [32]. So without loss of generality, in the rest of this paper, we assume that $\Lambda > v$ and $p_{\min} \leq v$.

Theorem 1. *For the scheduling problem under consideration, the worst-case bound of algorithm LPT$_v$ is 4/3 and the bound is tight.*

Proof. Let $\pi^{(v)}$ and π^* be the schedule generated by algorithm LPT$_v$ and an optimal schedule, respectively. Let $C_{\max}^{(v)}$ and $C_{\max}^*$ be the makespans of $\pi^{(v)}$ and π^*, respectively.

It is easy to see that

$$C^*_{\max} \geq \Lambda + r. \tag{1}$$

For schedule $\pi^{(v)}$, let λ be the total processing time of the jobs processed before the maintenance activity, let $C_i^{(v)}$ be the completion time of job J_i, and let J_m be the job that completes the last, i.e., $C_m^{(v)} = C_{\max}^{(v)}$. Recall that $\Lambda > v$. So job J_m must be scheduled to be processed after the maintenance in schedule $\pi^{(v)}$. According to the definition of job J_m, we know that jobs J_{m+1}, J_{m+2}, ..., J_n (if any) are scheduled to be processed before the maintenance activity in schedule $\pi^{(v)}$.

Now, consider the following two cases.

Case 1: $\lambda \geq s$. It is easy to see that, for schedule $\pi^{(v)}$, the maintenance activity starts just after the last job that is scheduled to be processed before it and hence there is no machine idle time before the maintenance. Note that the total processing time of the jobs that are scheduled after the maintenance in schedule $\pi^{(v)}$ is $\Lambda - \lambda$. Hence, we have

$$C_{\max}^{(v)} = \lambda + r + (\Lambda - \lambda).$$

That is

$$C_{\max}^{(v)} = \Lambda + r.$$

Combining this with (1), we have

$$C^*_{\max} \geq C_{\max}^{(v)},$$

which implies that schedule $\pi^{(v)}$ is also an optimal schedule.

Case 2: $\lambda < s$. It is easy to see that the maintenance activity starts at time s in schedule $\pi^{(v)}$ according to algorithm LPT_v. Now, consider the following three subcases.

Subcase 2.1: $m = 1$. It is easy to see that, in this subcase, job J_1 is the only job that is scheduled to be processed after the maintenance activity in schedule $\pi^{(v)}$. Clearly, we have $p_1 > v$ since job J_1 is not scheduled to be processed before time v in schedule $\pi^{(v)}$. According to the definition of job J_m, we know that jobs J_2, J_3, ..., J_n (if any) are scheduled to be processed before the maintenance in schedule $\pi^{(v)}$. Recall that the maintenance activity starts at time s in $\pi^{(v)}$. Hence, we have

$$C_{\max}^{(v)} = s + r + p_1. \tag{2}$$

Recall that $p_1 > v$. So job J_1 must be scheduled to be processed after the maintenance activity in any optimal schedule. Note that the maintenance activity can not start before time s. So we have

$$C^*_{\max} \geq s + r + p_1.$$

Combining this with (2), we have

$$C^*_{\max} \geq C_{\max}^{(v)},$$

which implies that schedule $\pi^{(v)}$ is also an optimal schedule.

Subcase 2.2: $m = 2$. In this case, there are at most two jobs processed after the maintenance activity in schedule $\pi^{(v)}$.

If $p_1 > v$, then we have $p_2 > v$, otherwise job J_2 should be scheduled to be processed before the maintenance activity in schedule $\pi^{(v)}$. So, if $p_1 > v$, we have

$$C_{\max}^{(v)} = s + r + p_1 + p_2. \tag{3}$$

Note that jobs J_1 and J_2 must be processed after the maintenance activity in any optimal schedule under the same assumption. Hence we have

$$C_{\max}^* \geq s + r + p_1 + p_2.$$

Combining this with (3), we have

$$C_{\max}^* \geq C_{\max}^{(v)},$$

which implies that schedule $\pi^{(v)}$ is also an optimal schedule.

If $p_1 \leq v$, it is easy to see that job J_2 is the only job that is scheduled to be processed after the maintenance activity in schedule $\pi^{(v)}$. So we have

$$C_{\max}^{(v)} = s + r + p_2. \tag{4}$$

Note that $p_1 + p_2 > v$ for that job J_2 is scheduled to be processed after time v in schedule $\pi^{(v)}$. Hence, at least one of the jobs J_1 and J_2 must be scheduled to be processed after the maintenance activity in any optimal schedule under the assumption of $p_1 \leq v$. Therefore, if $p_1 \leq v$, we have

$$C_{\max}^* \geq s + r + \min\{p_1, p_2\}.$$

Note that $p_1 \geq p_2$. So we have

$$C_{\max}^* \geq s + r + p_2.$$

Combining this with (4), we have

$$C_{\max}^* \geq C_{\max}^{(v)},$$

which implies that schedule $\pi^{(v)}$ is also an optimal schedule.

Case 2.3: $m \geq 3$. Recall that job J_m is scheduled to be processed after the maintenance in schedule π^v. So we have $s - \lambda < p_m$. Note that

$$C_{\max}^{(v)} = \lambda + (s - \lambda) + r + (\Lambda - \lambda).$$

So we have

$$C_{\max}^{(v)} < \Lambda + r + p_m. \tag{5}$$

Note that $p_1 \geq p_2 \geq \ldots \geq p_m$. So we have $p_m \leq \Lambda/m$. Combining this with (5), we have

$$C_{\max}^{(v)} < \Lambda + r + \Lambda/m.$$

That is

$$C_{\max}^{(v)} < (1 + 1/m)\Lambda + r.$$

Note that $r \geq 0$. So we have

$$C_{\max}^{(v)} < (1 + 1/m)(\Lambda + r).$$

Combining this with (1), we have

$$C_{\max}^{(v)} < (1 + 1/m)C_{\max}^*.$$

Recall that $m \geq 3$. So we have

$$C_{\max}^{(v)} < (4/3)C_{\max}^*.$$

Now, we have completed the proof that the worst-case bound of the LPT_v algorithm cannot be greater than 4/3. To show that this bound is tight, consider the following instance: $n = 3$, $p_1 = a + 2$, $p_2 = p_3 = a$, $s = 2a$, $t = 2a + 2$, and $r = 1$ with $a > 2$. According to the LPT_v algorithm, job J_1 is scheduled to be processed before the maintenance activity and the rest two jobs are scheduled to be processed after the maintenance activity which starts at time s. It is easy to see that $C_{\max}^{(v)} = 4a + 1$. However, an optimal schedule may schedule jobs J_2 and J_3 before the maintenance activity and job J_1 after the maintenance activity which also starts at time s. It is easy to see that the optimal makespan $C_{\max}^* = 3a + 3$. Hence, $C_{\max}/C_{\max}^* = (4a + 1)/(3a + 3) \to 4/3$ as $a \to \infty$.

This completes the proof.

3   Initial Start Time Sensitivity Analysis

One may have noticed that algorithm LPT_v sets time v, which is the latest start time of the maintenance activity, as the initial start time of the maintenance activity. It naturally invokes an interesting question that what about taking some other value in $[s, v)$ as the initial start time? The main aim of this section is trying to answer this question.

Now, let w be a number in $[s, v)$. By similar arguments as in algorithm LPT_v, we have algorithm LPT_w, which is the same as algorithm LPT_v except that every v is replaced by w in the statement of the algorithm. The other notations are defined analogously.

Theorem 2. *For any fixed $w \in [s, v)$, the worst-case bound of algorithm LPT_w is 2 and the bound is tight.*

Proof. It is easy to see that

$$C_{\max}^{(w)} \leq s + r + \Lambda.$$

Recall that $s \leq v < \Lambda$. So we have

$$C_{\max}^{(w)} < r + 2\Lambda.$$

Note that
$$C^*_{\max} \geq r + \Lambda.$$

So we have $C^{(w)}_{\max} < 2C^*_{\max}$.

To show that the bound is tight, consider the following instance: $n = 2$, $s = a$, $t = a + b + r$, $r = 1$ with $b < a$. Let $w = a + c$, where $c \in [0, b)$. Let $\epsilon = (b - c)/2$, $p_1 = a + c + \epsilon$, and $p_2 = b - c$. According to algorithm LPT$_w$, job J_1 is scheduled to be processed after the maintenance activity and job J_2 before the maintenance activity which starts at time s. The corresponding makespan $C^{(w)}_{\max} = 2a + 1 + c + \epsilon$. However, the optimal schedule arranges job J_1 before the maintenance activity and job J_2 after the maintenance activity which starts at time p_1. It is easy to check that the optimal makespan $C^*_{\max} = a + 1 + b + \epsilon$. Hence, $C^{(w)}_{\max}/C^*_{\max} = (2a + 1 + c + \epsilon)/(a + 1 + b + \epsilon) \to 2$ as $a \to \infty$.

This completes the proof.

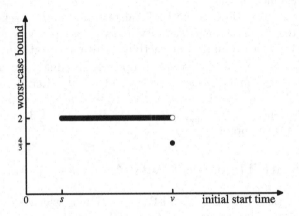

Fig. 1. The phenomenon of bound jump

Theorem 2 shows that the initial start time of the maintenance activity has no influence on the worst-case bound of the algorithm if it is taken from $[s, v)$ for that it is always 2 where s and v are the earliest and latest start times of the maintenance activity, respectively. However, according to Theorem 1, this bound jumps down to 4/3 if v is selected as the initial start time of the maintenance activity (see Fig. 1). This implies that the algorithm proposed by Yang et al. [32] is the best among this series of algorithms from the worst-case bound point of view. This is very interesting for that a slightly small increasing of the initial start time of the maintenance activity may incur such an unexpected improvement on the worst-case bound. This is also very important for real production for that the choice of an appropriate initial value may decrease the deviation of the objective from the optimal one sharply and hence may raise productivity. To the best of our knowledge, such an interesting phenomenon of *bound jump* has not been reported in the scheduling community.

References

1. Ben Abdellafou, K., Hadda, H., Korbaa, O.: An improved Tabu search meta-heuristic approach for solving scheduling problem with non-availability constraints. Arab. J. Sci. Eng. **44**(4), 3369–3379 (2019). https://doi.org/10.1007/s13369-018-3525-3
2. Baker, K.R., Trietsch, D.: Principles of Sequencing and Scheduling, 2nd edn. Wiley, Hoboken (2019)
3. Beaton, C., Diallo, C., Gunn, E.: Makespan minimization for parallel machine scheduling of semi-resumable and non-resumable jobs with multiple availability constraints. INFOR Inf. Syst. Oper. Res. **54**, 305–316 (2016). https://doi.org/10.1080/03155986.2016.1166795
4. Błażewicz, J., Ecker, K.H., Pesch, E., Schmidt, G., Weglarz, J.: Handbook on Scheduling: From Theory to Applications. Springer, Berlin (2007). https://doi.org/10.1007/978-3-540-32220-7
5. Breit, J.: Improved approximation for non-preemptive single machine flow-time scheduling with an availability constraint. Eur. J. Oper. Res. **183**, 516–524 (2007). https://doi.org/10.1016/j.ejor.2006.10.005
6. Chen, J.S.: Scheduling of nonresumable jobs and flexible maintenance activities on a single machine to minimize makespan. Eur. J. Oper. Res. **190**(1), 90–102 (2008). https://doi.org/10.1016/j.ejor.2007.06.029
7. Cheng, T.C.E., Wang, G.: An improved heuristic for two-machine flowshop scheduling with an availability constraint. Oper. Res. Lett. **26**, 223–229 (2000). https://doi.org/10.1016/S0167-6377(00)00033-X
8. Detti, P., Nicosia, G., Pacifici, A., de Lara, G.Z.M.: Robust single machine scheduling with a flexible maintenance activity. Comput. Oper. Res. **107**, 19–31 (2019). https://doi.org/10.1016/j.cor.2019.03.001
9. Hnaien, F., Yalaoui, F., Mhadhbi, A.: Makespan minimization on a two-machine flowshop with an availability constraint on the first machine. Int. J. Prod. Econ. **164**, 95–104 (2015). https://doi.org/10.1016/j.ijpe.2015.02.025
10. Huo, Y., Zhao, H.: Two machine scheduling subject to arbitrary machine availability constraint. Omega Int. J. Manage. Sci. **76**, 128–136 (2018). https://doi.org/10.1016/j.omega.2017.05.004
11. Kacem, I., Kellerer, H.: Approximation schemes for minimizing the maximum lateness on a single machine with release times under non-availability or deadline constraints. Algorithmica **80**, 3825–3843 (2018). https://doi.org/10.1007/s00453-018-0417-6
12. Kacem, I.: Levner, E: An improved approximation scheme for scheduling a maintenance and proportional deteriorating jobs. J. Ind. Mang. Optim. **12**, 811–817 (2016). https://doi.org/10.3934/jimo.2016.12.811
13. Kellerer, H., Pferschy, U., Pisinger, D.: Knapsack Problems. Springer, Berlin (2004). https://doi.org/10.1007/978-3-540-24777-7
14. Lee, C.Y.: Machine scheduling with an availability constraint. J. Glob. Optim. **9**, 395–416 (1996). https://doi.org/10.1007/BF00121681
15. Li, G., Liu, M., Sethi, S.P., Xu, D.: Parallel-machine scheduling with machine-dependent maintenance periodic recycles. Int. J. Prod. Econ. **186**, 1–7 (2017). https://doi.org/10.1016/j.ijpe.2017.01.014
16. Low, C., Li, R.K., Wu, G.H., Huang, C.L.: Minimizing the sum of absolute deviations under a common due date for a single-machine scheduling problem with availability constraints. J. Ind. Prod. Eng. **32**, 204–217 (2015). https://doi.org/10.1080/21681015.2015.1031196

17. Luo, W., Chen, L.: Approximation schemes for scheduling a maintenance and linear deteriorating jobs. J. Ind. Manag. Optim. **8**, 271–283 (2012). https://doi.org/10.3934/jimo.2012.8.271

18. Luo, W., Cheng, T.C.E., Ji, M.: Single-machine scheduling with a variable maintenance activity. Comput. Ind. Eng. **79**, 168–174 (2015). https://doi.org/10.1016/j.cie.2014.11.002

19. Luo, W., Ji, M.: Scheduling a variable maintenance and linear deteriorating jobs on a single machine. Inform. Process. Lett. **115**, 33–39 (2015). https://doi.org/10.1016/j.ipl.2014.08.011

20. Luo, W., Liu, F.: On single-machine scheduling with workload-dependent maintenance duration. Omega Int. J. Manage. Sci. **68**, 119–122 (2017). https://doi.org/10.1016/j.omega.2016.06.008

21. Mosheiov, G., Sarig, A.: Scheduling a maintenance activity to minimize total weighted completion-time. Comput. Math. Appl. **57**, 619–623 (2009). https://doi.org/10.1016/j.camwa.2008.11.008

22. Perez-Gonzalez, P., Framinan, J.M.: Single machine scheduling with periodic machine availability. Comput. Ind. Eng. **123**, 180–188 (2018). https://doi.org/10.1016/j.cie.2018.06.025

23. Pinedo, M.: Scheduling: Theory, Algorithms, and Systems, 5th edn. Springer, Cham (2016). https://doi.org/10.1007/978-3-319-26580-3

24. Wang, T., Baldacci, R., Lim, A., Hu, Q.: A branch-and-price algorithm for scheduling of deteriorating jobs and flexible periodic maintenance on a single machine. Eur. J. Oper. Res. **271**, 826–838 (2018). https://doi.org/10.1016/j.ejor.2018.05.050

25. Xu, D., Cheng, Z., Yin, Y., Li, H.: Makespan minimization for two parallel machines scheduling with a periodic availability constraint. Comput. Oper. Res. **36**, 1809–1812 (2009). https://doi.org/10.1016/j.cor.2008.05.001

26. Xu, D., Sun, K., Li, H.: Parallel machine scheduling with almost periodic maintenance and non-preemptive jobs to minimize makespan. Comput. Oper. Res. **35**, 1344–1349 (2008). https://doi.org/10.1016/j.cor.2006.08.015

27. Xu, D., Wan, L., Liu, A., Yang, D.-L.: Single machine total completion time scheduling problem with workload-dependent maintenance duration. Omega Int. J. Manage. Sci. **52**, 101–106 (2015). https://doi.org/10.1016/j.omega.2014.11.002

28. Xu, Z., Xu, D.: Single-machine scheduling with preemptive jobs and workload-dependent maintenance durations. Oper. Res. **15**(3), 423–436 (2015). https://doi.org/10.1007/s12351-015-0179-8

29. Xu, Z., Xu, D.: Single-machine scheduling with workload-dependent tool change durations and equal processing time jobs to minimize total completion time. J. Sched. **21**(4), 461–482 (2018). https://doi.org/10.1007/s10951-017-0543-z

30. Xu, D., Yang, D.-L.: Makespan minimization for two parallel machines scheduling with a periodic availability constraint: mathematical programming model, average-case analysis, and anomalies. Appl. Math. Model. **37**, 7561–7567 (2013). https://doi.org/10.1016/j.apm.2013.03.001

31. Xu, D., Yin, Y., Li, H.: Scheduling jobs under increasing linear machine maintenance time. J. Sched. **13**, 443–449 (2010). https://doi.org/10.1007/s10951-010-0182-0

32. Yang, D.-L., Hung, C.-L., Hsu, C.-J., Chern, M.-S.: Minimizing the makespan in a single machine scheduling problem with a flexible maintenance. J. Chin. Inst. Ind. Eng. **19**, 63–66 (2002). https://doi.org/10.1080/10170660209509183

Minimizing Energy on Homogeneous Processors with Shared Memory

Vincent Chau[1], Chi Kit Ken Fong[2], Shengxin Liu[3(✉)], Elaine Yinling Wang[1,4], and Yong Zhang[1]

[1] Shenzhen Institutes of Advanced Technology, Chinese Academy of Sciences, Shenzhen, China
[2] Chu Hai College of Higher Education, Hong Kong, China
[3] Nanyang Technological University, Singapore, Singapore
millersxliu@gmail.com
[4] School of Mathematical Sciences, Dalian University of Technology, Dalian, China

Abstract. Energy efficiency is a crucial desideratum in the design of computer systems, from small-sized mobile devices with limited battery to large scale data centers. In such computing systems, processor and memory are considered as two major power consumers among all the system components. One recent trend to reduce power consumption is using shared memory in multi-core systems, such architecture has become ubiquitous nowadays. However, implementing the energy-efficient methods to the multi-core processor and the shared memory separately is not trivial. In this work, we consider the energy-efficient task scheduling problem, which coordinates the power consumption of both the multi-core processor and the shared memory, especially focus on the general situation in which the number of tasks is more than the number of cores. We devise an approximation algorithm with guaranteed performance in the multiple cores system. We tackle the problem by first presenting an optimal algorithm when the assignment of tasks to cores is given. Then we propose an approximation assignment for the general task scheduling.

Keywords: Energy · Scheduling · Shared memory · Approximation algorithm

1 Introduction

Over the past decades, energy consumption is one of the major concerns in computer science. This attracts the attention of researchers in designing algorithms for minimizing energy consumption without performance degradation, which is referred to as energy efficiency problems.

One of the technologies used for energy efficiency is speed-scaling, where the processors are capable of varying its speed dynamically (or *Dynamic Voltage Scaling* (DVS)). The faster the processors run, the more energy they consume.

This work is supported by the CAS President's International Fellowship Initiative No 2020FYT0002, 2018PT0004.

M. Li (Ed.): FAW 2020, LNCS 12340, pp. 83–95, 2020.
https://doi.org/10.1007/978-3-030-59901-0_8

The idea is to complete all the given tasks before their deadlines with the slowest possible speed to minimize energy usage. The energy minimization problem of scheduling n tasks with release times and deadlines on a single-core processor that can vary its speed dynamically where preemption is allowed has been first studied in the seminal paper by Yao et al. [16].

Fu et al. [11] were the first to study the problem of minimizing the scheduling problem with the consideration of both the multi-core processor power and the shared main memory power. The challenge of this problem lies in balancing these two conflicting power consumption: executing tasks at a lower speed leads to a lower processor power consumption but a higher memory power consumption. They focus on the case where the number of cores is unbounded, and tasks are available at the beginning.

In this paper, we consider the energy-efficient task scheduling problem for multiple homogeneous cores and the main memory shared by these cores. Our objective is to find a feasible schedule with minimum energy consumption. We concentrate on the scenario when the number of tasks is more than the number of cores. Fu et al. [11] already pointed out that the setting is NP-hard. Hence, we aim to design an approximation algorithm to obtain a near-optimal performance with a theoretical guarantee. We have made the following technical contributions in this paper:

- For completeness, we present the optimal algorithm when there is a single-core.
- We extend the intuition of the single-core case to the multi-core case, where the problem is tackled in two steps:
  - We first present an optimal polynomial-time algorithm when the assignment of tasks to cores is given;
  - We propose an algorithm to assign tasks to cores, and show that it is a constant approximation algorithm.

Related Works. As a standard method to reduce the power consumption of the processor, DVS works by properly scaling the voltage of the processor and has been widely utilized for a single-core processor in recent decades. Yao et al. [16] were the first to give a polynomial-time algorithm to compute an optimal schedule. The time complexity has been further improved in [13] and [14]. More works along this line can be found in the surveys [1,2]. Albers et al. [3] proved the NP-hardness of the multi-core DVS problem when tasks could not migrate[1] and gave several approximation algorithms for various special cases. More recently, there also exist some works focusing on scheduling tasks at an appropriate speed to create an idle period in which the processor can be switched into the sleep state [5,6,9,12,15].

Besides speed-scaling, another way of reducing energy consumption is dynamic resource sleep (DRS) management, which powers off the machines when the machines do not have many tasks to process. For instance, the storage cluster in the data centers can be turned off to save energy during low utilization period.

---

[1] A task must be scheduled on a single core.

The min-gap strategy is one of the approaches in DRS. When the machines are idle, they are transited to the suspended state without any energy consumption. However, a small amount of energy will be consumed in the process of waking up the machines from the suspended state. Hence, the objective of the min-gap strategy is to find a schedule such that the number of idle periods can be minimized. This problem has been first studied by Baptiste [7], and has been improved in [8]. More works along this line with various setting can be found in [4, 10].

The work most related to ours is from [11]. They studied the problem of task scheduling for multiple cores with shared memory. However, they only studied the case when the number of cores is larger than the number of tasks. They presented optimal solutions for different task models, such as all tasks are available at the beginning, or for the case where tasks have agreeable deadline, i.e., later release time implies later deadline. Finally, they tested the proposed algorithms with simulations under the condition when the system is reasonably loaded. However, if the system is overloaded (in particular when there are more tasks than the number of available cores), their algorithm cannot be used, and therefore, the primary motivation of this paper is to address this problem.

Paper Organization. The rest of the paper is organized as follows. Section 2 describes the system model and presents the problem definition studied in this paper. In Sects. 3 and 4, we consider the single-core and multi-core cases, respectively.

2   Preliminaries

In this section, we present the system and task models studied throughout this paper. Then we formally define our problem based on these models.

2.1   System and Task Model

We consider the energy-efficient task scheduling problem for multiple homogeneous cores and the main memory shared by these cores. We assume that each core has an individual voltage supply and the speed of each core changes in a continuous fashion. We present both definitions of the multi-core processor power and the shared main memory power as below. The power consumption of each core [12] is measured as $P(s) = \gamma + \beta s^{\alpha}$, where $\alpha > 1$, β are hardware-specified constants [16], and γ denotes the static power of the core and βs^{α} denotes the dynamic power. When $\gamma > 0$, it means that the cores consume energy even when they are not executing a task; Thus, these cores can be turned into the sleep state for energy saving. When a core is running at speed s for $t > 0$ units time, it can perform a total workload of $s \cdot t$ and consumes $t \cdot P(s)$ energy.

For the shared main memory power consumption, we consider the static power of the memory. That is, as long as one of the cores executes a task, the shared memory needs to be in the active state, which consumes γ_m energy per unit of time. We refer to the *static energy* when the energy consumption is due to the active state of the core or of the shared memory, while the *dynamic energy* refers to the energy consumption due to the execution speed of the cores.

We have a set of n tasks, $\{1, 2, \ldots, n\}$, where each task i is associated with a release time r_i, a deadline d_i, and a non-negative workload w_i. Without loss of generality, we suppose that tasks are sorted in non-decreasing order of their deadlines, i.e., $d_1 \leq d_2 \ldots \leq d_n$. A schedule is *feasible* if and only if all tasks are completed by their deadlines. In this work, we consider the case where tasks are available from the beginning, i.e., $r_i = 0, \forall i$. We also let $W_{i,j} = \sum_{k=i}^{j} w_k$. We use *makespan* to represent the maximum completion time, i.e., the last moment the cores are executing a task.

2.2 Problem Definition

In this paper, we consider the following problem. Given a set of n tasks that need to be scheduled on a system, as described above, where the system is associated with p cores and a shared memory. Our objective is to find a schedule with the minimum energy consumption such that the tasks are scheduled on a unique core, and such that they are completed before their deadline. Thus, we aim to minimize the following function:

$$E = \sum_{i=1}^{p} \left(\gamma c_i + \beta \int_0^{d_n} s_i(t)^\alpha \, dt \right) + \gamma_m \max_{i=1}^{p} \{c_i\} \tag{1}$$

where $s_i(t)$ is the running speed of the core i at time t, c_i is the completion time of the core i.

3   Warm-Up: Single Core Case

This section studies the single-core case in which we propose an algorithm that computes the optimal schedule. We first consider the situation where $\gamma_m = 0$, and we will discuss the case of $\gamma_m > 0$ subsequently. The proposed algorithm is a consequence of [6]. They studied the problem of task scheduling with a single speed-scalable core, with static energy and when tasks have agreeable deadline, i.e., $r_j \leq r_i \Leftrightarrow d_j \leq d_i$. The running complexity time of the algorithm is $O(n^3)$. We show in that when all tasks are released at time 0, and the running time becomes $O(n^2)$. The algorithm in [6] is based on the algorithm in [16] that computes the minimum energy consumption schedule by only considering the *dynamic energy*; In each iteration, we find the *maximum* intensity interval of tasks, which is the minimum speed such that the set of selected tasks must be scheduled in order to be completed before their respective deadlines. Such a speed can be calculated by dividing the sum of the workload of the tasks over the length of the interval. Finally, we remove the considered interval from the instance/schedule, and we repeat this procedure until all tasks are scheduled. In our case, the *static energy* has to be taken into account. We observe that, since the energy function $P(s)/s$ is convex, there exists a unique speed s^* such that a task with any other speed $s' \neq s^*$ will consume more energy. Formally, we have the following definition:

Definition 1 ([6,12]). *For the single core case, the critical speed is defined as:* $s^\star = \arg\min_s \frac{P(s)}{s} = \sqrt[\alpha]{\frac{\gamma}{\beta(\alpha-1)}}.$

Since tasks are available at time 0, the considered interval at each step will be in the following form $[0, d_j]$ for some task j. After removing this interval, the next interval will be in the following form $[d_j, d_i]$ for some tasks $j < i$. Since at each step, we find the *maximum* intensity interval of tasks, it means that the future intervals will have lower intensity (task execution speed). We then have the following proposition.

Proposition 1. *The schedule returned by Algorithm 1 has a non-increasing running speed.*

The idea is to consider the tasks in groups where the speed of a group is calculated as the minimum speed for these tasks to ensure that they can be completed before their deadlines. According to Definition 1, tasks should be scheduled at speed s^* if it is possible. In particular, the strategy of speed selection is as follows :

- When the speed of a group of tasks is less than s^*, we speed up the schedule in order to minimize the energy consumption.
- When the speed is larger than s^*, it means that it is not possible to decrease the speed (to s^*) because it will violate the deadline constraints.

By combining the algorithms from [6,16] and the observation of Definition 1, we obtain Algorithm 1. After initializing in lines 1–2, we compute the critical speed of s^* in line 3. Then we schedule the tasks in the while-loops (lines 4–13). For each iteration, we compute a group of consecutive tasks starting from the current task T_i with minimum speed s such that no deadlines are violated in lines 5–6. Then we schedule this group of tasks at speed s^* if $s \leq s^*$ (line 8); s, otherwise (line 10). The iteration keeps looping until we have scheduled all the tasks. The analysis of Algorithm 1 is shown as follows.

Theorem 1. *Algorithm 1 computes the optimal schedule in $O(n^2)$ time.*

This algorithm is a direct consequence of the algorithm in [6]. We remark that Algorithm 1 can be applied to the case of $\gamma_m > 0$ by changing $s^* \leftarrow \arg\min_s (P(s) + \gamma_m)/s$ in line 2. The optimality and the running time of the algorithm remain unchanged.

4   Multi-core Case

In this section, we propose a polynomial-time approximation algorithm with a bounded performance guarantee. We first present an optimal algorithm when the task assignment is given in Sect. 4.1, then we propose an assignment algorithm that generates an approximate solution with a bounded ratio in Sect. 4.2.

Algorithm 1. Optimal algorithm for single core and common release time tasks

---

1: Set $t \leftarrow 0$ and $i \leftarrow 1$.
2: Sort the set of tasks in non-decreasing order of their deadline
3: Compute the critical speed $s^\star \leftarrow \arg\min_s P(s)/s$.
4: **while** $i \leq n$ **do**
5:     Compute $W_{i,j} \leftarrow \sum_{k=i}^{j} w_k, \forall i \leq j \leq n$.
6:     Compute $j^\star \leftarrow \arg\max_{i \leq j \leq n} \frac{W_{i,j}}{d_j - t}$ and $s \leftarrow \frac{W_{i,j^\star}}{d_{j^\star} - t}$.
7:     **if** $s \leq s^\star$ **then**
8:         Schedule tasks $\{i, \ldots, j^\star\}$ at speed $s^\star$ until $d_{j^\star}$.
9:     **else**
10:         Schedule tasks $\{i, \ldots, j^\star\}$ at speed s until $d_{j^\star}$.
11:     **end if**
12:     Set $t \leftarrow d_{j^\star}$ and $i \leftarrow j^\star + 1$.
13: **end while**

4.1 Computing Optimal Schedule for a Given Task Assignment

According to the assignment of the tasks, we compute the minimum energy consumption of each core with Algorithm 1. We aim to calculate the optimal length of the sleeping time of the shared memory, which is the length between the end of the schedule and the maximum deadline. We then divide the schedule into several intervals, which are delimited by different relevant completion times of each core. The rationale of this approach is that the energy consumption function is convex in the function of the memory sleeping time between two critical consecutive time points, which may not be the case overall. Thus, for our case, it is also crucial to find such a set of critical time points for separating the feasible domain of the memory sleeping time. We first consider the single-core case and observe the behavior of energy consumption when we increase the memory sleeping time.

Given a schedule on core i, by Proposition 1, let $s_1 > s_2 > \ldots > s_{k_i}$ be the set of different speeds. We also let $\mathcal{Z}_k$ be the set of tasks that are scheduled at speed s_k, i.e., $j \in \mathcal{Z}_k$ if and only if task j is scheduled at speed s_k. We refer to the set of tasks running at the same speed as a *block of tasks*. Because of the convexity of the power function, each task is scheduled at a constant speed. Thus there are at most n different speeds in an optimal solution, and in particular, there are at most n blocks of tasks.

A first observation is that when we increase the length of the sleeping time, only the execution speed of the last block of tasks increases. This comes from the fact that the last block has the slowest speed among all blocks of tasks, and increasing the execution speed of another block of tasks will incur a higher energy consumption. Therefore, we only need to find out how far we can increase the sleeping time until the last block's speed reaches the second last block's speed. When the execution speed of the last block is equal to the speed of the second last block, they are merged into a new block. See Fig. 1 for an illustration of the different completion times. In the first schedule, we increase the speed of the last block of tasks until the execution speed is equal to the execution speed of the second last block of tasks (with single hatching). We can obtain the completion time e_2 at this moment. Similarly, we can obtain the completion time e_1 by changing the execution speed of the second last and the last blocks of tasks.

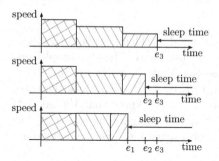

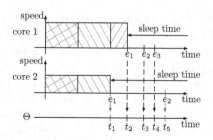

Fig. 1. Illustration of different ending time when we modify the speed of the last block of tasks. From the first schedule, we increase the execution speed of the last block until it has the same execution speed as the previous block, and we get the second schedule as well as the completion time e_2.

Fig. 2. The set Θ is the union of different completion times of 2 cores.

Thus, we can compute the different completion time in the following way: let e_{k-u} be the completion time of the schedule if we only use the speeds in $\{s_1, s_2, \ldots, s_{k-u}\}$ and we have:

$$e_{k-u} = \frac{\sum_{j=k-u}^{k} \sum_{i \in \mathcal{Z}_j} w_i}{s_{k-u}} + \sum_{j=1}^{k-u-1} \frac{\sum_{i \in \mathcal{Z}_j} w_i}{s_j} \quad \text{where } 1 \leq u \leq k. \quad (2)$$

We apply the same principle to the multiple cores case. See Fig. 2 for an example on two cores.

Definition 2. *Let Θ be the set of relevant completion times on all cores.*

$$\Theta = \bigcup_{i=1}^{p} \{e_{n_i-u} \mid 1 \leq u \leq n_i\}$$

At each step, we increase the speed of the last block of tasks. If there are n_i tasks on core i, then such a modification occurs at most $n_i - 1$ times which creates $n_i - 1$ new completion time. Thus adding the original completion time, we have n_i for core i, which implies that the set Θ contains at most n completion time. Let $t_1 < t_2 < \ldots < t_{|\Theta|}$ be the time in Θ. We analyze each zone of Θ which is delimited by two consecutive time. In any interval $[t_\ell, t_{\ell+1}]$, the time t_ℓ

corresponds to either an actual completion time of a core, or the time when one block's speed reaches another block's speed (see Fig. 1). Then, the cost of the schedule is a convex function depends on the makespan.

Proposition 1. *The cost of the schedule is convex depending on the makespan for each interval $[t_\ell, t_{\ell+1}]$ where $1 \leq \ell \leq |\Theta|$.*

Proof. To simplify the notation, let $B_{i,k}$ be the total workload of the k-th block of tasks on core i, k_i be the number of blocks on core i and c_i be the completion time of core i. Without loss of generality, assume that the cores are sorted in non-decreasing order of completion time, i.e., $c_1 \leq c_2 \leq \ldots \leq c_p$. Since the core p has the largest completion time among all cores, we know that the shared memory will be active until c_p. Moreover, let $s_{i,b}$ be the running speed of the b-th block on core i. The energy consumption of a schedule S can be defined as follows: $E(S) = \sum_{i=1}^{p} \sum_{b=1}^{k_i} \frac{B_{i,b}}{s_{i,b}} P(s_{i,b}) + \gamma_m c_p$.

Let δ_ℓ denote the length of sleeping time of the new schedule, i.e., the period between the end of the schedule and the end of the zone ℓ. For a given interval $[t_\ell, t_{\ell+1}]$, we can separate the cost into two parts: the first part is a constant according to the makespan, while the other part depends on the makespan. We denote p' as the number of cores whose makespan is less than $c_p = t_{\ell+1}$. The energy consumption of the first p' cores is not affected when the maximum makespan is in $[t_\ell, t_{\ell+1}]$. In fact, only the last block on cores $p'+1, \ldots, p$ will be affected by the sleeping time δ_ℓ (see Fig. 3).

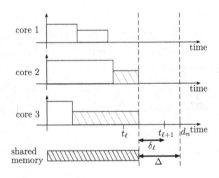

Fig. 3. Illustration of how the sleeping time δ_ℓ affects the schedule. Only the blocks of tasks with hatching are affected by the sleeping time.

Therefore, the speed of the last block for each core will be increased when the length of sleeping time δ_ℓ increases. More formally, the new speed will be $s_{i,k_i}(\delta_\ell) = \frac{B_{i,k_i}}{\frac{B_{i,k_i}}{s_{i,k_i}} - \delta_\ell}$. We are now able to define the energy cost function depending on $\delta_\ell \in [0, t_{\ell+1} - t_\ell]$ as follows:

Algorithm 2. Optimal schedule for a given task assignment

Require: Assignment of tasks to cores.
1: **for** each core i **do**
2:    Apply Algorithm 1 and obtain schedule S.
3: **end for**
4: Compute different completion time on each core as defined in Eq. (2) which are then collected in Θ.
5: Compute the optimal solution $E_\ell(S)$ for each interval $[t_\ell, t_{\ell+1}]$ s.t. $t_\ell, t_{\ell+1} \in \Theta$.
6: **return** $\min_{t_\ell \in \Theta} E_\ell(S)$

$$E_\ell(S, \delta_\ell) = \sum_{i=1}^{p'} \sum_{b=1}^{k_i} \frac{B_{i,b}}{s_{i,b}} P(s_{i,b}) + \sum_{i=p'+1}^{p} \sum_{b=1}^{k_i-1} \frac{B_{i,b}}{s_{i,b}} P(s_{i,b})$$
$$+ \sum_{i=p'+1}^{p} \left(\frac{B_{i,k_i}}{s_{i,k_i}} - \delta_\ell \right) P(s_{i,k_i}(\delta_\ell)) + \gamma_m(c_p - \delta_\ell).$$

Since the first part does not depend on δ_ℓ, it is a constant. We only need to show that the second part of the cost function is convex on δ_ℓ. In particular, we can show that the second derivative is positive (omitted due to space constraint).

From the construction of the completion time in Θ, we know that the starting time of a block is not in $[t_\ell, t_{\ell+1}]$. We show this by contradiction. Suppose that there exists a core such that the last block of tasks starts at a time t in $[t_\ell, t_{\ell+1}]$. When we increase the execution speed of the last block of tasks, the completion time e cannot be before t. However, there exists a time $t_\ell < t < e < t_{\ell+1}$ such that the execution speed of the last block of tasks is equal to the execution speed of the second last block of tasks. We have a contradiction because e should have been computed and added to Θ. Thus, we have $B_{i,k_i} - s_{i,k_i} \delta_\ell \geq 0$ when $\delta_\ell \in [0, t_{\ell+1} - t_\ell]$ which leads to a feasible solution. Thus, the function $E_\ell(S, \delta_\ell)$ is convex. $\square$

Let $E_\ell(S)$ be the optimal solution when the makespan of the schedule is in $[t_\ell, t_{\ell+1}]$. Then, we have $E_\ell(S) = \min_{\delta_\ell \in [0, t_{\ell+1} - t_\ell]} E_\ell(S, \delta_\ell)$, $\forall t_\ell, t_{\ell+1} \in \Theta$. Combining everything, our optimal algorithm for this part is described in Algorithm 2. In particular, we apply Algorithm 1 for the single-core case to each core in lines 1–3. Then we compute the completion time for each core in line 4, followed by the computation of the optimal solution in each interval, as shown in Proposition 1. Finally, we return the optimal schedule. The analysis of Algorithm 2 is listed as follows.

Theorem 2. *Given an assignment of tasks to cores, Algorithm 2 computes the optimal solution in $O(n^2 p)$ time.*

4.2 Task Assignment

As shown in Sect. 4.1, we can compute the schedule with the minimum energy consumption if the assignment of tasks is given. We now present a task assignment scheme that produces a bounded approximate solution.

Lower Bound. The intuition of our lower bound is to allow scheduling the same task on different cores at the same time. Since we relax a constraint of the problem, the energy consumption of such a schedule must be at least the optimal energy consumption of our problem. In particular, by relaxing this constraint, the workload of the cores will be balanced, and the execution speed of the cores are identical at any time, so the energy consumption will be lower because of the convexity of the power function.

Approximation Algorithm. Our approximation algorithm, as shown in Algorithm 3, is based on a simple task assignment method. Initially, we sort tasks in non-decreasing order of their deadlines in line 1. Then, we assign the task to the core with the lowest load (i.e., the core with the least total assigned workload) one by one in lines 2–4. Note that this assignment is similar to the one proposed by Albers et al. in [3]. Finally, we apply Algorithm 2 on this assignment and generate the corresponding schedule in line 5. In the rest of this section, we will prove that Algorithm 3 has a constant approximation ratio.

Algorithm 3. Approximation Algorithm

Require: Set of tasks J
 1: Sort tasks in non-decreasing order of deadline.
 2: **for** each task $j \in J$ **do**
 3:    Assign task j to the core with the lowest load.
 4: **end for**
 5: Apply Algorithm 2 on the assignment of tasks to compute minimal energy consumption.

First, we need to show that there is an upper bound of the speed of the cores when scheduling the last block of tasks.

Proposition 2. *The execution speed of a core at the end of the schedule (the last block of tasks) such that any tasks of this block does not end at its deadline, is at most* $s_m^\star = \sqrt[\alpha]{\frac{\gamma + \gamma_m}{\beta(\alpha - 1)}}$.

Proof. We prove this claim by contradiction. Suppose we have an optimal schedule, and the last block of tasks of a core has an execution speed, which is strictly higher than $s_m^\star$ such that no task of this block ends at its deadline. By considering this particular core and the shared memory, we can decrease the total energy consumption by decreasing the execution speed to $s_m^\star$ or until a task of this block reaches a deadline, which decreases the energy consumption. Thus we have a contradiction with the fact that the schedule is optimal.   □

Theorem 3. *Algorithm 3 has an approximation ratio of* $\max\left\{1 + \frac{\gamma_m}{\gamma}, 2^{\alpha+2}\right\}$.

Proof. Let LB (resp. AMS) be the lower bound of the schedule (resp. the schedule returned by the algorithm in [3]). Noted that in AMS, the static energy is not taken into account, so the execution speeds of the tasks are as slow as possible.

Let us consider the assignment of tasks returned by Algorithm 3 (lines 1–4). We construct a schedule S with different execution speeds for each task according to this assignment and show that its cost can be bounded. We recall that the energy consumption of Algorithm 3 is no more than the energy consumption of S, since Algorithm 3 has the minimum energy consumption for the same assignment of tasks. The schedule S is constructed as follows. For each task, we choose the maximum execution speed between AMS and LB. Since AMS is a feasible schedule, by scheduling tasks faster, the resulting schedule is still feasible. The proof is divided into two parts: **Part (a)** is on the *static energy* while **Part (b)** is devoted to the cost induced by the *dynamic energy*. Then, it is obvious that the approximation ratio is bounded by the ratios obtained by either the ratio in static energy or the one in dynamic energy.

Part (a). By construction, the execution speed of tasks in S is higher or equal to the execution speed of the same task in LB, then the static energy of the cores in S is at most the static energy of the cores in LB. Similarly, the energy consumption of the shared memory in S is at most γ_m/γ times the static energy of the cores in LB. It is because the makespan is at most the total running time of all cores, so by multiplying γ_m/γ (we schedule some tasks with speed $s_m^\star$ if the initial speed was in $[s^\star, s_m^\star]$), we obtain an upper bound of the energy cost of the shared memory from Proposition 2. More formally, we have the following equations: $S_{core}^{(s)} \le LB_{core}^{(s)}$ and $S_{mem}^{(s)} \le \frac{\gamma_m}{\gamma} LB_{core}^{(s)}$, where $S_{core}^{(s)}$ (resp. $S_{mem}^{(s)}$) is the energy consumption of the static energy of the cores (resp. shared memory) in S, and $LB_{core}^{(s)}$ and $LB_{mem}^{(s)}$ are defined similarly. Thus $LB_{core}^{(s)} + LB_{mem}^{(s)} \le S_{core}^{(s)} + S_{mem}^{(s)} \le LB_{core}^{(s)} + \frac{\gamma_m}{\gamma} LB_{core}^{(s)}$. And the approximation ratio is $\frac{LB_{core}^{(s)} + \frac{\gamma_m}{\gamma} LB_{core}^{(s)}}{LB_{core}^{(s)} + LB_{mem}^{(s)}} \le \frac{LB_{core}^{(s)}(1 + \frac{\gamma_m}{\gamma})}{LB_{core}^{(s)}} = 1 + \frac{\gamma_m}{\gamma}$.

Since the optimal static energy is at least $LB_{core}^{(s)} + LB_{mem}^{(s)}$, we have that the approximation ratio of the static energy is $(1 + \frac{\gamma_m}{\gamma})$.

Part (b). On the other hand, we study the *dynamic energy* of the schedule S, i.e., the energy consumption induced by the execution speed of the cores. Clearly, we have $AMS \le S^{(d)}$ and $LB^{(d)} \le S^{(d)}$ by construction.

We also know that the dynamic energy consumption is at most the sum of dynamic energy consumptions of AMS and LB. That is, $S^{(d)} \le AMS + LB^{(d)}$. Let $LB_{AMS}^{(d)}$ be the lower bound of the dynamic energy of AMS. Albers et al. [3] shows that $LB_{AMS}^{(d)} \le AMS \le 2^{\alpha+1} LB_{AMS}^{(d)}$. So we have the following equations: $LB^{(d)} \le S^{(d)} \le 2^{\lambda+1} LB_{AMS}^{(d)} + LB^{(d)}$ and $LB_{AMS}^{(d)} \le S^{(d)} \le 2^{\alpha+1} LB_{AMS}^{(d)} + LB^{(d)}$ By summing these two equations, we have:

$$LB^{(d)} + LB^{(d)}_{AMS} \leq 22\, S^{(d)} \leq 2\left(2^{\alpha+1} LB^{(d)}_{AMS} + LB^{(d)}\right)$$

$$\frac{LB^{(d)} + LB^{(d)}_{AMS}}{2} \leq S^{(d)} \leq 2^{\alpha+1} LB^{(d)}_{AMS} + LB^{(d)}$$

Since the optimal dynamic energy is at least $\frac{LB^{(d)}+LB^{(d)}_{AMS}}{2}$, the approximation ratio of the dynamic energy is $\frac{2^{\alpha+1} LB^{(d)}_{AMS}+LB^{(d)}}{\frac{LB^{(d)}+LB^{(d)}_{AMS}}{2}} \leq 2^{\alpha+2}$.

As the total energy consumption consists of both static and dynamic energy costs, combining the inequalities (3) and (4), the approximation ratio of Algorithm 3 is $\max\left\{1 + \frac{\gamma_m}{\gamma}, 2^{\alpha+2}\right\}$.   □

References

1. Albers, S.: Energy-efficient algorithms. Commun. ACM **53**(5), 86–96 (2010)
2. Albers, S.: Algorithms for dynamic speed scaling. In: 28th International Symposium on Theoretical Aspects of Computer Science, STACS 2011. LIPIcs, vol. 9, pp. 1–11 (2011)
3. Albers, S., Müller, F., Schmelzer, S.: Speed scaling on parallel processors. Algorithmica **68**(2), 404–425 (2014). https://doi.org/10.1007/s00453-012-9678-7
4. Angel, E., Bampis, E., Chau, V.: Low complexity scheduling algorithms minimizing the energy for tasks with agreeable deadlines. Discrete Appl. Math. **175**, 1–10 (2014)
5. Antoniadis, A., Huang, C.C., Ott, S.: A fully polynomial-time approximation scheme for speed scaling with sleep state. In: Proceedings of the Annual ACM-SIAM Symposium on Discrete Algorithms (SODA) (2015)
6. Bampis, E., Dürr, C., Kacem, F., Milis, I.: Speed scaling with power down scheduling for agreeable deadlines. Sustain.Comput. Inform. Syst. **2**(4), 184–189 (2012)
7. Baptiste, P.: Scheduling unit tasks to minimize the number of idle periods: a polynomial time algorithm for offline dynamic power management. In: Proceedings of the 17th Annual ACM-SIAM SODA, 2006. pp. 364–367. ACM Press (2006)
8. Baptiste, P., Chrobak, M., Dürr, C.: Polynomial-time algorithms for minimum energy scheduling. ACM Trans. Algorithms **8**(3), 26:1–26:29 (2012)
9. Chen, J.J., Hsu, H.R., Kuo, T.W.: Leakage-aware energy-efficient scheduling of real-time tasks in multiprocessor systems. In: Proceedings of IEEE Real-Time and Embedded Technology and Applications Symposium (RTAS), pp. 408–417 (2006)
10. Demaine, E.D., Ghodsi, M., Hajiaghayi, M., Sayedi-Roshkhar, A.S., Zadimoghaddam, M.: Scheduling to minimize gaps and power consumption. J. Sched. **16**(2), 151–160 (2013). https://doi.org/10.1007/s10951-012-0309-6
11. Fu, C., Chau, V., Li, M., Xue, C.J.: Race to idle or not: balancing the memory sleep time with DVS for energy minimization. J. Comb. Optim. **35**(3), 860–894 (2018). https://doi.org/10.1007/s10878-017-0229-7
12. Irani, S., Shukla, S., Gupta, R.: Algorithms for power savings. ACM Trans. Algorithms **3**(4), 41:1–41:23 (2007)
13. Li, M., Yao, A.C., Yao, F.F.: Discrete and continuous min-energy schedules for variable voltage processors. Proc. Natl. Acad. Sci. USA **103**(11), 3983–3987 (2006)

14. Li, M., Yao, F.F., Yuan, H.: An $O(n^2)$ algorithm for computing optimal continuous voltage schedules. In: Gopal, T.V., Jäger, G., Steila, S. (eds.) TAMC 2017. LNCS, vol. 10185, pp. 389–400. Springer, Cham (2017). https://doi.org/10.1007/978-3-319-55911-7_28

15. Wu, W., Li, M., Wang, K., Huang, H., Chen, E.: Speed scaling problems with memory/cache consideration. J. Sched. **21**(6), 633–646 (2018). https://doi.org/10.1007/s10951-018-0565-1

16. Yao, F., Demers, A., Shenker, S.: A scheduling model for reduced CPU energy. In: Proceedings of IEEE Annual Symposium on Foundations of Computer Science (FOCS), pp. 374–382 (1995)

Approximation Schemes for Subset Sum Ratio Problems

Nikolaos Melissinos[1](✉), Aris Pagourtzis[2], and Theofilos Triommatis[3]

[1] Université Paris-Dauphine, PSL University, CNRS, LAMSADE,
75016 Paris, France
nikolaos.melissinos@dauphine.eu
[2] School of Electrical and Computer Engineering, National Technical University of
Athens, Polytechnioupoli, 15780 Zografou, Athens, Greece
pagour@cs.ntua.gr
[3] School of Electrical Engineering, Electronics and Computer Science, University of
Liverpool, Liverpool L69-3BX, UK
Theofilos.Triommatis@liverpool.ac.uk

Abstract. We consider the Subset Sum Ratio Problem (SSR), in which given a set of integers the goal is to find two subsets such that the ratio of their sums is as close to 1 as possible, and introduce a family of variations that capture additional meaningful requirements. Our main contribution is a generic framework that yields fully polynomial time approximation schemes (FPTAS) for problems in this family that meet certain conditions. We use our framework to design explicit FPTASs for two such problems, namely *Two-Set Subset-Sum Ratio* and *Factor-r Subset-Sum Ratio*, with running time $\mathcal{O}(n^4/\varepsilon)$, which coincides with the best known running time for the original SSR problem [15].

Keywords: Approximation scheme · Subset-sums ratio · Knapsack problems · Combinatorial optimization

1 Introduction

Subset sum computations are of key importance in computing, as they appear either as standalone tasks or as subproblems in a vast amount of theoretical and practical methods coping with important computational challenges. As most of subset sum problems are NP-hard, an effort was made over the years to come up with systematic ways of deriving approximation schemes for such problems. Important contributions in this direction include works by Horowitz and Sahni [9,10], Ibarra and Kim [11], Sahni [20], Woeginger [22] and Woeginger and Pruhs [19]. Inspired by these works we define and study families of variations of the Subset Sum Ratio Problem which is a combinatorial optimization problem introduced and shown NP-hard by Woeginger and Yu [23]. The formal definition of the problem is as follows:

© Springer Nature Switzerland AG 2020
M. Li (Ed.): FAW 2020, LNCS 12340, pp. 96–107, 2020.
https://doi.org/10.1007/978-3-030-59901-0_9

Subset-Sums Ratio Problem (SSR). *Given a set* $A = \{a_1, \ldots, a_n\}$ *of* n *positive integers, find two nonempty and disjoint sets* $S_1, S_2 \subseteq \{1, \ldots, n\}$ *that minimize the ratio*

$$\frac{\max\{\sum_{i \in S_1} a_i, \sum_{j \in S_2} a_j\}}{\min\{\sum_{i \in S_1} a_i, \sum_{j \in S_2} a_j\}}$$

One of our motives to study SSR stems from the fact that it is the optimization version of the decision problem Equal Subset Sum (ESS) which is related to various other concepts and problems as we explain below. ESS is defined as follows:

Equal Sum Subsets Problem (ESS). *Given a set* $A = \{a_1, \ldots, a_n\}$ *of* n *positive integers, are there two nonempty and disjoint sets* $S_1, S_2 \subseteq \{1, \ldots, n\}$ *such that* $\sum_{i \in S_1} a_i = \sum_{j \in S_2} a_j$?

Even if this problem has been in the literature for many years, it is still being studied with a recent work begin [16]. Variations of this problem have been studied and proven NP-hard by Cieliebak *et al.* in [6,7], where pseudo - polynomial time algorithms were also presented for many of these problems. ESS is a fundamental problem appearing in many scientific areas. For example, it is related to the Partial Digest problem that comes from molecular biology [3,4], to allocation mechanisms [14], to tournament construction [13], to a variation of the Subset Sum problem, namely the Multiple Integrated Sets SSP, which finds applications in cryptography [21]. A restricted version of ESS, namely when the sum of the input values is strictly less than $2^n - 1$ is guaranteed to have a solution, however it is not known how to find it; this version belongs to the complexity class PPP [18].

The first FPTAS for SSR was introduced by Bazgan *et al.* in [1] and more recently a simpler but slower FPTAS was introduced in [17] and a faster one in [15]; the latter is the fastest known so far for the problem. Variations of ESS were studied and shown NP-hard in [5–7], where also pseudo - polynomial time algorithms were presented for some of them and it was left open whether the corresponding optimization problems admit an FPTAS. Here we address that question in the affirmative for two of those problems (namely for *Equal Sum Subsets From Two Sets* [5][1] and for *Factor-r Sum Subsets* [7]) and provide a framework that can be potentially used to give an FPTAS for most of the remaining ones, if not for all, as well as for many other subset sum ratio problems.

Let us note that, for the exemplary problems that we study here there may exist more efficient approximation algorithms, e.g. by using techniques such as those in [2,12] for knapsack, however it is not clear if and how such techniques can be adapted in a generic way to take into account the additional restrictions that are captured by our framework. Moreover, our primary goal is to provide an as generic as possible framework to cope with such problems, at the cost of sacrificing optimality in efficiency.

---

[1] It is not hard to show that the optimization version of *Equal Sum Subsets From Two Sets* can be reduced to *Two-Set Subset-Sum Ratio* for which we provide an FPTAS here.

Our results and organization of the paper are as follows. In Sect. 2 we define two families of variations of SSR problems that are able to capture additional restrictions. Our main result, presented in Sect. 3, is a method to obtain an FPTAS for any problem in these families the definition of which meets certain conditions. In the last two sections we use our framework to present FPTASs for two variations of SSR, namely 2-Set SSR and Factor-r Subset Sum Ratio; to the best of our knowledge, no approximation algorithm was known so far for these problems.

2   Families of Variations of SSR

In this section we shall define two families of variations of the SSR problem. In [15], the function $\mathcal{R}(S_1, S_2, A)$ was defined as:

Definition 1 (Ratio of two subsets). *Given a set $A = \{a_1, \ldots, a_n\}$ of n positive integers and two sets $S_1, S_2 \subseteq \{1, \ldots, n\}$ we define $\mathcal{R}(S_1, S_2, A)$ as follows:*

$$\mathcal{R}(S_1, S_2, A) = \begin{cases} 0 & \text{if } S_1 = \emptyset \text{ and } S_2 \neq \emptyset \\ \frac{\sum_{i \in S_1} a_i}{\sum_{i \in S_2} a_i} & \text{if } S_2 \neq \emptyset, \\ +\infty & \text{otherwise.} \end{cases}$$

Here we will also define and use $\mathcal{MR}(S_1, \ldots, S_k, A)$ which is a generalization of $\mathcal{R}(S_1, S_2, A)$ to $k > 2$ sets:

Definition 2 (Max ratio of k subsets). *Given a set $A = \{a_1, \ldots, a_n\}$ of n positive integers and two sets $S_1, \ldots, S_k \subseteq \{1, \ldots, n\}$ we define:*

$$\mathcal{MR}(S_1, \ldots, S_k, A) = \max\{\mathcal{R}(S_i, S_j, A) \mid i \neq j \text{ and } i, j \in \{1, \ldots, k\}\}$$

In order to keep our expressions as simple as possible we will use the above functions throughout the whole paper.

Let us now define the first family of variations of SSR. We want this family to contain as many problems as possible. In a general case we may not have just a set of numbers as our input but a graph or that has a weights on the edges or the vertices. For such reasons we will use the following notation.

Family of Subset-Sum Ratio Problems (F-SSR). *A problem $\mathcal{P}$ in F-SSR is a combinatorial optimization problem $(\mathcal{I}, k, \mathcal{F})$ where:*

- *$\mathcal{I}$ is a set of instances each of which is a pair (E, w) where $E = \{e_1, \ldots, e_n\}$ is a set of ground elements and $w : E \mapsto \mathbb{R}^+$ is a weight function which maps every element e_i to a positive number a_i;*
- *k defines the number of subsets of $\{1, \ldots, n\}$ we are searching for;*
- *$\mathcal{F}$ gives the set of feasible solutions as follows: for any input (E, w), $\mathcal{F}(k, E)$ is a collection of k-tuples of nonempty and disjoint subsets of $\{1, \ldots, n\}$, and given $(k, E, (S_1, \ldots, S_k))$ we can check in polynomial time whether $(S_1, \ldots, S_k) \in \mathcal{F}(k, E)$.*

The goal of $\mathcal{P}$ *is to find for an instance* (E, w) *a feasible solution* $(S_1^*, \ldots, S_k^*)$ *such that*

$$\mathcal{MR}(S_1^*, \ldots, S_k^*, A) = \min\{\mathcal{MR}(S_1, \ldots, S_k, A) \mid (S_1, \ldots, S_k) \in \mathcal{F}(k, E)\}$$

where $A = \{a_i = w(e_i) \mid e_i \in E\}$

Remark 1. Note that under this definition of $F\text{-}SSR$ the function w of an instance (E, w) does not play any role in deciding whether a k-tuple $(S_1, \ldots, S_k)$ is feasible or infeasible solution; in other words, the element weights do not affect feasibility, only their indices do. Consequently, for a specific problem $\mathcal{P} = (\mathcal{I}, k, \mathcal{F}) \in F\text{-}SSR$ and two different instances (E, w) and (E, w') in $\mathcal{I}$ with the same ground elements E, the feasible solutions of the two instances are the same.

We will now introduce a family that is similar to $F\text{-}SSR$ but there is a major difference which is an extra condition. In this family we know (we give it as input), the minimum between the maximum values of the solution. This is rather technical and it will become obvious in the following paragraphs.

Family of Semi - Restricted Subset-Sum Ratio Problems (*Semi-Restr icted F-SSR*). *For every problem* $\mathcal{P} = (\mathcal{I}, k, \mathcal{F})$ *in F-SSR, we define an associated optimization problem* $\mathcal{P}' = (\mathcal{I}', k', \mathcal{F}')$ *as follows:*

– *the set of instances of* $\mathcal{P}'$ *is*

$$\mathcal{I}' = \{(E, w, m) \mid (E, w) \in \mathcal{I} \text{ and } m \in \{1, \ldots |E|\}\}$$

– $k' = k$
– *the collection of feasible solutions of instance* $(E, w, m) \in \mathcal{I}'$ *is given by:*

$$\mathcal{F}'(k, E, w, m) = \{(S_1, \ldots, S_k) \in \mathcal{F}(k, E) \mid \min_{j \in \{1, \ldots, k\}} \{\max_{i \in S_j} w(e_i)\} = w(e_m)\}$$

and the goal of $\mathcal{P}'$ *is to find for an instance* (E, w, m) *a feasible solution* $(S_1^*, \ldots, S_k^*)$ *such that*

$$\mathcal{MR}(S_1^*, \ldots, S_k^*, A) = \min\{\mathcal{MR}(S_1, \ldots, S_k, A) \mid (S_1, \ldots, S_k) \in \mathcal{F}(k, E)\}$$

where $A = \{a_i = w(e_i) \mid e_i \in E\}$. *We define the family of problems Semi-Restricted F-SSR as the class of problems* $\{\mathcal{P}' \mid \mathcal{P} \in F\text{-}SSR\}$.

Remark 2. We note that if a problem belongs to *Semi-Restricted F-SSR* it cannot belong to $F\text{-}SSR$ because there is the extra condition for a solution $(S_1, \ldots, S_k)$ to be feasible, $\min_{j \in \{1, \ldots k\}} \{\max_{i \in S_j} w(e_i)\} = w(e_m)$ which depends on the weight function w and not only on the set of elements E as is the case for problems in $F\text{-}SSR$.

Remark 3. It is obvious that if there exists a deterministic polynomial time Turing Machine that can decide if a solution is feasible for a problem $\mathcal{P}$ in F-SRR then we can construct another deterministic polynomial time Turing Machine that takes into account the extra condition to decide, if a solution is feasible for the semi restricted version $\mathcal{P}'$ in Semi Restricted F-SRR of the previous.

We must note that F-SSR contains many problems of many from different areas in computer science and could prove useful to get an FPTAS for them if we could develop a pseudo - polynomial algorithm with a particular property (it will be explained in the next section) for the semi restricted versions of them. This family includes subset sum ratio problems with matroid restrictions, graph restrictions, cardinality restrictions (including partition problems). Some more problems could be scheduling problems and knapsack.

To give some examples, we will present some problems that belong in F-SSR. For the first two, the proof that they actually do belong in F-SSR will be presented in Sect. 4 and Sect. 5 respectively. We must note that the decision version of $Factor$-r SSR that follows was studied in [7]. For these two problems, we will introduce FPTAS algorithms in the following sections. Moreover we will present other problems of F-SSR that have more complicated constraints and could prove interesting to be studied in the future.

Two-Set Subset-Sum Ratio Problem (2-Set SSR). *Let* $A = \{(a_1, b_1),$ $\dots, (a_n, b_n)\}$ *be a set of pairs of real numbers. We are searching for two nonempty and disjoint sets* $S_1, S_2 \subseteq \{1, ..., n\}$ *that minimize*

$$\frac{\max\{\sum_{i \in S_1} a_i, \sum_{j \in S_2} b_j\}}{\min\{\sum_{i \in S_1} a_i, \sum_{j \in S_2} b_j\}}.$$

Factor-r Subset-Sum Ratio Problem (Factor-r SSR). *Given a set* $A = \{a_1, \dots, a_n\}$ *of n positive integers and a real number $r \geq 1$, find two nonempty and disjoint sets* $S_1, S_2 \subseteq \{1, \dots, n\}$ *that minimize the ratio*

$$\frac{\max\{r \cdot \sum_{i \in S_1} a_i, \sum_{j \in S_2} a_j\}}{\min\{r \cdot \sum_{i \in S_1} a_i, \sum_{j \in S_2} a_j\}}.$$

In [8] there were introduced digraph constraints for the subset sum problem which can easily be modeled via our framework. Generally we are able demand as constraints of S_1 and S_2 to be a specific property considering the vertices of the graph, for example we may demand that the solution consists of independent sets or dominant sets etc. Not only can we impose constraints for the sets of vertices but we can impose constraints on the edges of the graph as well. Finally we may impose restrictions that concern both edges and vertices at the same time, for example take into account vertices that form a complete graph.

The same way we define the constraints from a graph we are able to demand that the solution of a problem consists of independent sets of a given matroid.

Subset-Sum Ratio with Matroid Constraints. *Given a matroid* $\mathcal{M}(E, I)$ *and a weight function* $w : E \to \mathbb{R}^+$. *We want to find two non empty and non equal sets* $S_1, S_2 \in I$ *such that:*

$$\frac{\max\{\sum_{x \in S_1} w(x), \sum_{y \in S_2} w(y)\}}{\min\{\sum_{x \in S_1} w(x), \sum_{y \in S_2} w(y)\}}.$$

Before we continue to the next section we will present two lemmas which give us information about the solutions which are feasible for both problems in F-SSR and $Semi$-$Restricted$ F-SSR. Moreover we must note that all the proofs for the theorems and the lemmas are deferred to the full version of the paper.

Lemma 1. *Let $\mathcal{P} = (\mathcal{I}, k, \mathcal{F})$ a problem in F-SSR and $\mathcal{P}' = (\mathcal{I}', k', \mathcal{F}')$ the semi restricted version of $\mathcal{P}$. If $(E, w) \in \mathcal{I}$ and $(E, w', m) \in \mathcal{I}'$ are the instances of $\mathcal{P}$ and $\mathcal{P}'$ respectively then any feasible solution $(S_1, \ldots, S_k)$ of the instance (E, w', m) of $\mathcal{P}'$ is also a feasible solution of the instance (E, w) of $\mathcal{P}$.*

Lemma 2. *Let $\mathcal{P} = (\mathcal{I}, k, \mathcal{F})$ a problem in F-SSR and $\mathcal{P}' = (\mathcal{I}', k', \mathcal{F}')$ the semi restricted version of $\mathcal{P}$. If E is a set of elements and w, w' two weight functions such that:*

$$\text{For any } i, j \in \{1, \ldots n\}, \quad w(e_i) < w(e_j) \Leftrightarrow w'(e_i) \leq w'(e_j)$$

then any feasible solution $(S_1, \ldots, S_k)$ for the instance (E, w) of $\mathcal{P}$ is a feasible solution for the instance (E, w', m) of $\mathcal{P}'$ if

$$w(e_m) = \min_{j \in \{1, \ldots, k\}} \{\max\{w(e_i) \mid i \in S_j\}\}$$

3 A Framework Yielding FPTAS for Problems in F-SSR

In the following theorem we want to define a scale parameter δ which we will use later to change the size of our input and the pseudo - polynomial algorithms will run in polynomial time. In addition by using these parameters, we will define the properties that the sets of the output should satisfy to be $(1 + \varepsilon)$ approximation. These parameters are not unique but any other number should to the trick as long as all the properties of the theorem bellow are satisfied.

Theorem 1. *Let $A = \{a_1, \ldots, a_n\}$ be a set of positive real numbers, $\varepsilon \in (0, 1)$, two sets $S_{1Opt}, S_{2Opt} \subseteq \{1, \ldots, n\}$ and any numbers w, m, δ that satisfy:*

- $0 < w \leq \min \left(\sum_{i \in S_{1Opt}} a_i, \sum_{i \in S_{2Opt}} a_i \right)$
- $n \geq m \geq \max (|S_{1Opt}|, |S_{2Opt}|)$,
- $\delta = (\varepsilon \cdot w)/(3 \cdot m)$

If $S_1, S_2 \subseteq \{1, \ldots, n\}$ are two non-empty sets such that:

- $w \leq \min \left(\sum_{i \in S_1} a_i, \sum_{i \in S_2} a_i \right)$
- $n \geq m \geq \max (|S_1|, |S_2|)$,
- $1 \leq MR(S_1, S_2, A') \leq MR(S_{1Opt}, S_{2Opt}, A')$ *where* $A' = \{\lfloor \frac{a_1}{\delta} \rfloor, \ldots, \lfloor \frac{a_n}{\delta} \rfloor\}$

Then the following inequality holds

$$1 \leq MR(S_1, S_2, A) \leq (1 + \varepsilon) \cdot MR(S_{1Opt}, S_{2Opt}, A) .$$

The next theorem presents the conditions that should be met to construct an FPTAS algorithm for a problem that belongs in $F\text{-}SSR$. Keep in mind that this framework should be considered similar to linear programming, i.e. if there is a way to prove that a problem belongs to $F\text{-}SSR$ and at the same time there is a pseudo - polynomial time algorithm for its semi-restricted version then we can obtain an FPTAS algorithm.

Theorem 2. *Let $\mathcal{P} = (\mathcal{I}, \mathcal{F}, \mathcal{M}, \mathcal{G})$ be a problem in F-SSR and $\mathcal{P}' = (\mathcal{I}', \mathcal{F}', \mathcal{M}, \mathcal{G})$ its corresponding problem in Semi Restricted F-SSR. If for problem $\mathcal{P}'$ there exists an algorithm that solves exactly all instances $A = \{a_1, \ldots a_n, m\} \in \mathcal{I}'$ in which all a_i values are integers in time $\mathcal{O}(poly(n, a_m))$, then $\mathcal{P}$ admits an FPTAS.*

Now we will present an algorithm that approximates $\mathcal{P}$ using the algorithm for $\mathcal{P}'$. We will denote the algorithm that returns the exact solution for $\mathcal{P}'$ by $\mathcal{SOL}_{ex,\mathcal{P}'}(A)$.

Algorithm 1. FPTAS for the problem $\mathcal{P}$ [$\mathcal{SOL}_{apx,\mathcal{P}}(A)$ function]

Input: A set $A = \{a_1, \ldots, a_n\}, a_i \in \mathbb{R}^+$.
Output: Sets with max ratio $(1 + \varepsilon)$ to the optimal max ratio for the problem $\mathcal{P}$.
1: $(S_1^*, \ldots, S_k^*) \leftarrow \{\emptyset, \ldots, \emptyset\}$
2: **for** $m \leftarrow 1$ **to** n **do**
3:      $\delta \leftarrow \frac{\varepsilon \cdot a_m}{3 \cdot n}$
4:      $A^{(m)} \leftarrow \emptyset$
5:      **for** $i \leftarrow 1$ **to** n **do**
6:          $a_i' \leftarrow \lfloor \frac{a_i}{\delta} \rfloor$
7:          $A^{(m)} \leftarrow A^{(m)} \cup \{a_i'\}$
8:      **end for**
9:      $A^{(m)} \leftarrow A^{(m)} \cup \{m\}$
10:     $(S_1', \ldots, S_k') \leftarrow \mathcal{SOL}_{ex,\mathcal{P}'}(A^{(m)})$
11:     **if** $\mathcal{MR}(S_1', \ldots, S_k', A) \leq \mathcal{MR}(S_1^*, \ldots, S_k^*, A)$ **then**
12:         $(S_1^*, \ldots, S_k^*) \leftarrow (S_1', \ldots, S_k')$
13:     **end if**
14: **end for**
15: **return** $(S_1^*, \ldots, S_k^*)$

In the next sections we will give some examples of how this framework works by using Theorem 2 to find an FPTAS algorithm for some problems.

4   2-Set SSR

Here, we will design an FPTAS algorithm for 2-*Set SSR*. We must note that faster approximation algorithms could be developed for this particular problem but this is not the scope of this section.

We will begin by proving that this problem belongs in F-SSR. We will match the 2-Set SSR with a problem $(\mathcal{I}, \mathcal{F}, \mathcal{M}, \mathcal{G})$ in F-SSR. If we let the set of instances $\mathcal{I}$ contain sets of positive numbers $A = \{a_1, \ldots, a_{2 \cdot n}\} = \{a_1, \ldots, a_n, b_1, \ldots, b_n\}$, and the set of feasible solutions $\mathcal{F}$ contain all the pairs of sets (S_1, S_2) such that $S_1 \subseteq \{1, \ldots, n\}$, $S_2 \subseteq \{n+1, \ldots, 2 \cdot n\}$, $\nexists$ (i,j) such that $i \in S_1, j \in S_2$ with $i \equiv j \pmod{n}$, the objective $\mathcal{M} = MR(S_1, S_2, A))$ and the goal function $\mathcal{G} = \min$, then the 2-Set SSR problem coincides with $(\mathcal{I}, \mathcal{F}, \mathcal{M}, \mathcal{G})$ which is a problem in F-SSR.

Now we shall present a pseudo - polynomial algorithm that finds an optimal solution for *Semi-Restricted* 2-*Set* SSR. Our algorithm employs two separate algorithms for two different cases.

Algorithm 2. *Semi-Restricted 2-Set SSR* solution [$SOL(A, m)$ function]

Input: a set $A = \{a_1, \ldots, a_{2 \cdot n}\}$, $a_i \in \mathbb{Z}^+$, and an integer m, $1 \leq m \leq 2 \cdot n$.
Output: the sets of an optimal solution for *Semi-Restricted 2-Set SSR*.
1: $S_1' \leftarrow \emptyset$, $S_2' \leftarrow \emptyset$, $S_{min} \leftarrow \emptyset$, $S_{max} \leftarrow \emptyset$
2: **if** $m \leq n$ **then**
3:     $(p, p') \leftarrow (0, n)$
4: **else if** $n < m \leq 2 \cdot n$ **then**
5:     $(p, p') \leftarrow (n, 0)$
6: **end if**
7: $S_{min} \leftarrow \{i \mid i \in \{1, \ldots, n\} \text{ and } a_{i+p} \leq a_m\} \setminus \{m - p\}$
8: $S_{max} \leftarrow \{i \mid i \in \{1, \ldots, n\} \text{ and } a_{i+p'} \geq a_m\} \setminus \{m - p + p'\}$
9: **if** $S_{max} \neq \emptyset$ **then**
10:     $(S_1, S_2) \leftarrow SOL_{Case1}(A, m, S_{min}, S_{max})$
11:     $(S_1', S_2') \leftarrow SOL_{Case2}(A, m, S_{min}, S_{max})$
12:     **if** $MR(S_1, S_2, A) < MR(S_1', S_2', A)$ **then**
13:         $(S_1', S_2') \leftarrow (S_1, S_2)$
14:     **end if**
15: **end if**
16: **return** S_1', S_2'

We will continue with the presentation of algorithms $SOL_{Case1}(A, m, S_{min}, S_{max})$ and $SOL_{Case2}(A, m, S_{min}, S_{max})$. Let us first define a function that will simplify their presentation.

Definition 3 (LTST: Larger Total Sum Tuple selection). *Given two tuples* $\vec{v_1} = (S_1, S_2, x)$ *and* $\vec{v_2} = (S_1', S_2', x')$ *we define the function* $\mathsf{LTST}(\vec{v_1}, \vec{v_2})$ *as follows:*

$$\mathsf{LTST}(\vec{v_1}, \vec{v_2}) = \begin{cases} \vec{v_2} & \text{if } \vec{v_1} = (\emptyset, \emptyset, 0) \text{ or } x' > x, \\ \vec{v_1} & \text{otherwise} . \end{cases}$$

We will use this function to compare the sum of the sets $S_1 \cup S_2$ and $S_1' \cup S_2'$ i.e.

$$x = \sum_{i \in S_1 \cup S_2} a_i \qquad \text{and} \qquad x' = \sum_{i \in S_1' \cup S_2'} a_i$$

In the next algorithm we study the case 1. In case 1 we consider that we need to use an element that its weight is greater than the sum of the elements' weights that could belong to the other set. In this case the set with the largest total weight contains only one element and the other set contains all the allowed elements (elements that have no conflicts).

Algorithm 3. Case 1 solution [$\mathcal{SOL}_{Case1}(A, m, S_{min}, S_{max})$ function]

Input: a set $A = \{a_1, \ldots, a_{2 \cdot n}\}$, $a_i \in \mathbb{Z}^+$ and an integer m, $1 \leq m \leq 2 \cdot n$ and $S_{min}, S_{max} \subseteq \{1, \ldots, n\}$.

Output: Case 1 optimal solution for *Semi-Restricted 2-Set SSR*.

1: $S_1' \leftarrow \emptyset$, $S_2' \leftarrow \emptyset$
2: **if** $m \leq n$ **then**
3:     $p \leftarrow 0$, $p' \leftarrow n$
4: **else**
5:     $p \leftarrow n$, $p' \leftarrow 0$
6: **end if**
7: $Q \leftarrow a_m + \sum_{i \in S_{min}} a_{i+p}$
8: **for all** $i \in S_{max}$ and $a_{i+p'} > Q$ **do**
9:     $a \leftarrow 0$
10:    **if** $i \in S_{min}$ **then**
11:        $a \leftarrow a_{i+p}$
12:    **end if**
13:    **if** $a_{i+p'}/(Q - a) < \mathcal{MR}(S_1', S_2', A)$ **then**
14:        $S \leftarrow \{j + p \mid j \in S_{min}$ or $j = m - p\} \smallsetminus \{i + p\}$
15:        $(S_1', S_2') \leftarrow (S, \{i + p'\})$
16:    **end if**
17: **end for**
18: **return** S_1', S_2'

In case 2 we consider that the largest (weighted) element doesn't necessarily dominate the sum of the weights of the second set. In this case we create a three dimensional matrix whose first dimension represents the elements we have already used, the second represents the difference of the sets' sums and the third is rather technical and it used to be sure that we won't overwrite tuples that have wanted properties. In the cells we store the two sets of indices and the total sum of their weights. Moreover when the third dimension has the value 1 then this means that these sets could be a part of a feasible solution.

Algorithm 4. Case 2 solution $[\mathcal{SOL}_{Case2}(A, m, S_{min}, S_{max})$ function]

Input: a set $A = \{a_1, \ldots, a_{2 \cdot n}\}$, $a_i \in \mathbb{Z}^+$ and an integer m, $1 \leq m \leq 2 \cdot n$ and $S_{min}, S_{max} \subseteq \{1, ..., n\}$.

Output: Case 2 optimal solution for *Semi-Restricted 2-Set SSR*.

1: $S'_1 \leftarrow \emptyset$, $S'_2 \leftarrow \emptyset$
2: **if** $m \leq n$ **then**
3:     $p \leftarrow 0$, $p' \leftarrow n$
4: **else**
5:     $p \leftarrow n$, $p' \leftarrow 0$
6: **end if**
7: $Q \leftarrow a_m + \sum_{i \in S_{min}} a_{i+p}$
8: $T[i, d, l] \leftarrow \{\emptyset, \emptyset, 0\}$, $\forall\, (i, d, l) \in \{0, \ldots, n\} \times \{-2 \cdot Q, \ldots, Q\} \times \{0, 1\}$
9: $T[0, a_m, 0] \leftarrow (\{m\}, \emptyset, a_m)$
10: **for** $i \leftarrow 1$ **to** n **do**
11:     **for all** $(d, l) \in \{-2 \cdot Q, \ldots, Q\} \times \{0, 1\}$
12:         $(S_1, S_2, x) \leftarrow T[i-1, d, l]$ **do**
13:         $T[i, d, l] \leftarrow \mathsf{LTST}(T[i, d, l], T[i-1, d, l])$
14:         $d' \leftarrow d + a_{i+p}$
15:         **if** $i \in S_{min}$ **then**
16:             $T[i, d', l] \leftarrow \mathsf{LTST}(T[i, d', l], (S_1 \cup \{i+p\}, S_2, x + a_{i+p}))$
17:         **end if**
18:         $d' \leftarrow d - a_{i+p'}$
19:         **if** $i \in S_{max}$ **and** $d' \geq -2 \cdot Q$ **then**
20:             $T[i, d', 1] \leftarrow \mathsf{LTST}(T[i, d', 1], (S_1, S_2 \cup \{i+p'\}, x + a_{i+p'}))$
21:         **else if** $i \notin S_{max}$ **and** $d' \geq -2 \cdot Q$ **then**
22:             $T[i, d', l] \leftarrow \mathsf{LTST}(T[i, d', l], (S_1, S_2 \cup \{i+p'\}, x + a_{i+p'}))$
23:         **end if**
24:     **end for**
25: **end for**
26: **for** $d \leftarrow -2 \cdot Q$ **to** Q **do**
27:     $(S_1, S_2, x) \leftarrow T[n, d, 1]$
28:     **if** $\mathcal{MR}(S_1, S_2, A) < \mathcal{MR}(S'_1, S'_2, A)$ **then**
29:         $S'_1 \leftarrow S_1$, $S'_2 \leftarrow S_2$
30:     **end if**
31: **end for**
32: **return** (S'_1, S'_2)

Theorem 3. *Algorithm 2 runs in time* $\mathcal{O}(n^2 \cdot a_m)$.

Since Algorithm 2 is a pseudo - polynomial time algorithm for the *Semi-Restricted 2-Set SSR* which solves the instances with integer values and runs in time $\mathcal{O}(poly(n, a_m))$, by using Theorem 2 we get that *2-Set SSR* admits an *FPTAS*. Furthermore, by using Algorithm 1 we have the following:

Theorem 4. *For 2-Set SSR and for every* $\varepsilon \in (0, 1)$ *we can find an* $(1 + \varepsilon)$ *approximation solution in time* $\mathcal{O}(n^4/\varepsilon)$.

5   Approximation of *SSR* and *Factor-r SSR*

In this section we will use the algorithm we design for the 2-*Set SSR* in order to approximate the original problem *SSR* and another one variation of *SSR*, the *Factor-r SSR*.

Before we approximate these problems we will prove that both of them are in *F-SSR*. Starting with the *SSR*, it is easy to identify it with a problem $(\mathcal{I}, \mathcal{F}, \mathcal{M}, \mathcal{G})$ in $F-SSR$: we let the set of instances $\mathcal{I}$ contain sets of positive integers $A = \{a_1, \ldots, a_n\}$, the set of feasible solutions $\mathcal{F}$ contain all the pairs of sets (S_1, S_2) such that $S_1 \cup S_2 \subseteq \{1, \ldots, n\}$, $S_1 \cap S_2 = \emptyset$, the measure be $\mathcal{M} = \mathcal{MR}(S_1, S_2, A)$, and the goal function be $\mathcal{G} = \min$.

Regarding *Factor-r SSR*, we identify it with a problem $(\mathcal{I}, \mathcal{F}, \mathcal{M}, \mathcal{G})$ in $F-SSR$, by letting the set of instances $\mathcal{I}$ contain sets of positive numbers $A = \{a_1, \ldots, a_{2n}\} = \{a_1, \ldots a_n, r \cdot a_1, \ldots, r \cdot a_n\}$ with $a_i \in \mathbb{Z}^+$ for $i \in \{1, \ldots, n\}$, $r \in \mathbb{R}$, the set of feasible solutions $\mathcal{F}$ contain all pairs of sets (S_1, S_2) such that $S_1 \subseteq \{1, \ldots, n\}$ and $S_2 \subseteq \{n+1, \ldots, 2n\}$ and $\forall\, (i, j), i \in S_1 \wedge j \in S_2 \Rightarrow i + n \neq j$, the measure be $\mathcal{M} = \mathcal{MR}(S_1, S_2, A))$, and the goal function be $\mathcal{G} = \min$.

For both problems we can modify their input in order to match the input of 2-*Set SSR*. Specifically, is not hard to see that an optimal solution for *SSR* with input $A = \{a_1, \ldots, a_n\}$ is an optimal solution for 2-*Set SSR* with input $A = \{\{a_1, a_1\} \ldots, (a_n, a_n)\}$ and vice versa. The same applies to an optimal solution for *Factor-r SSR* with input $(\{a_1, \ldots, a_n\}, r)$ and an optimal solution of 2-*Set SSR* with input $A = \{(a_1, r \cdot a_1) \ldots, (a_n, r \cdot a_n)\}$. Furthermore the feasible solutions for 2-*Set SSR*, with the specific input we discussed above, are the same with the ones for *SSR* (respectively for *Factor-r SSR*). So if we find an $(1 + \varepsilon)$ approximating solution for the 2-*Set SSR* problem with input $A = \{(a_1, a_1) \ldots, (a_n, a_n)\}$ (resp. with input $A = \{(a_1, r \cdot a_1) \ldots, (a_n, r \cdot a_n)\}$) then this is an $(1 + \varepsilon)$ approximating solution for *SSR* (resp. for *Factor-r SSR*).

References

1. Bazgan, C., Santha, M., Tuza, Z.: Efficient approximation algorithms for the subset-sums equality problem. J. Comput. Syst. Sci. **64**(2), 160–170 (2002). https://doi.org/10.1006/jcss.2001.1784
2. Chan, T.M.: Approximation schemes for 0-1 knapsack. In: SOSA 2018, pp. 5:1–5:12 (2018)
3. Cieliebak, M., Eidenbenz, S.: Measurement errors make the partial digest problem NP-hard. In: Farach-Colton, M. (ed.) LATIN 2004. LNCS, vol. 2976, pp. 379–390. Springer, Heidelberg (2004). https://doi.org/10.1007/978-3-540-24698-5_42
4. Cieliebak, M., Eidenbenz, S., Penna, P.: Noisy data make the partial digest problem NP-hard. In: Benson, G., Page, R.D.M. (eds.) WABI 2003. LNCS, vol. 2812, pp. 111–123. Springer, Heidelberg (2003). https://doi.org/10.1007/978-3-540-39763-2_9
5. Cieliebak, M., Eidenbenz, S., Pagourtzis, A., Schlude, K.: Equal sum subsets: complexity of variations. Technical report 370. ETH Zürich, Department of Computer Science (2003)

6. Cieliebak, M., Eidenbenz, S., Pagourtzis, A.: Composing equipotent teams. In: Lingas, A., Nilsson, B.J. (eds.) FCT 2003. LNCS, vol. 2751, pp. 98–108. Springer, Heidelberg (2003). https://doi.org/10.1007/978-3-540-45077-1_10

7. Cieliebak, M., Eidenbenz, S., Pagourtzis, A., Schlude, K.: On the complexity of variations of equal sum subsets. Nordic J. Comput. **14**(3), 151–172 (2008)

8. Gourvès, L., Monnot, J., Tlilane, L.: Subset sum problems with digraph constraints. J. Comb. Optim. **36**(3), 937–964 (2018). https://doi.org/10.1007/s10878-018-0262-1

9. Horowitz, E., Sahni, S.: Computing partitions with applications to the knapsack problem. J. ACM **21**(2), 277–292 (1974). https://doi.org/10.1145/321812.321823

10. Horowitz, E., Sahni, S.: Exact and approximate algorithms for scheduling nonidentical processors. J. ACM **23**(2), 317–327 (1976). https://doi.org/10.1145/321941.321951

11. Ibarra, O.H., Kim, C.E.: Fast approximation algorithms for the knapsack and sum of subset problems. J. ACM **22**(4), 463–468 (1975). https://doi.org/10.1145/321906.321909

12. Jin, C.: An improved FPTAS for 0-1 knapsack. ICALP **76**(1–76), 14 (2019)

13. Khan, M.A.: Some problems on graphs and arrangements of convex bodies. Ph.D. thesis, University of Calgary (2017). https://prism.ucalgary.ca/handle/11023/3765

14. Lipton, R.J., Markakis, E., Mossel, E., Saberi., A.: On approximately fair allocations of indivisible goods. In: Proceedings of the 5th ACM Conference on Electronic Commerce (EC 2004), New York, NY, USA, 17–20 May 2004, pp. 125–131 (2004)

15. Melissinos, N., Pagourtzis, A.: A faster FPTAS for the subset-sums ratio problem. In: Wang, L., Zhu, D. (eds.) COCOON 2018. LNCS, vol. 10976, pp. 602–614. Springer, Cham (2018). https://doi.org/10.1007/978-3-319-94776-1_50

16. Mucha, M., Nederlof, J., Pawlewicz, J., Wegrzycki, K.: Equal-subset-sum faster than the meet-in-the-middle. ESA **73**(1–73), 16 (2019)

17. Nanongkai, D.: Simple FPTAS for the subset-sums ratio problem. Inf. Process. Lett. **113**(19–21), 750–753 (2013)

18. Papadimitriou, C.H.: On the complexity of the parity argument and other inefficient proofs of existence. J. Comput. Syst. Sci. **48**, 498532 (1994)

19. Pruhs, K., Woeginger, G.J.: Approximation schemes for a class of subset selection problems. Theoret. Comput. Sci. **382**(2), 151–156 (2007). https://dblp.org/rec/bib/journals/tcs/PruhsW07

20. Sahni, S.: Algorithms for scheduling independent tasks. J. ACM **23**(1), 116–127 (1976). https://dblp.org/rec/bib/journals/jacm/Sahni76

21. Voloch, N.: MSSP for 2-D sets with unknown parameters and a cryptographic application. Contemp. Eng. Sci. **10**(19), 921–931 (2017)

22. Woeginger, G.J.: When does a dynamic programming formulation guarantee the existence of a fully polynomial time approximation scheme (FPTAS)? INFORMS J. Comput. **12**(1), 57–74 (2000). https://doi.org/10.1287/ijoc.12.1.57.11901

23. Woeginger, G.J., Yu, Z.: On the equal-subset-sum problem. Inf. Process. Lett. **42**(6), 299–302 (1992). https://doi.org/10.1016/0020-0190(92)90226-L

Two-Way Jumping Automata

Szilárd Zsolt Fazekas[(⊠)], Kaito Hoshi, and Akihiro Yamamura

Graduate School of Engineering Science, Akita University, 1-1 Tegatagakuenmachi,
Akita 010-8502, Japan
{szilard.fazekas,yamamura}@ie.akita-u.ac.jp, m8018308@s.akita-u.ac.jp

Abstract. The recently introduced one-way jumping automata are
strictly more powerful than classical finite automata (FA) while maintaining decidability in most of the important cases. We investigate the
extension of the new processing mode to two-way deterministic finite
automata (2DFA), resulting in deterministic finite automata which can
jump to the nearest letter they can read, with jumps allowed in either
direction. We show that two-way jumping automata are strictly more
powerful than one-way jumping ones and that alternative extensions of
2DFA with jumping mode lead to equivalent machines. We also prove
that the class of languages accepted by the new model is not closed under
the usual language operations. Finally we show how one could change
the model to terminate on every input.

Keywords: Finite automata · Two-way finite automata · One-way
jumping · Two-way jumping · Nonsequential processing

1 Introduction

Relatively recently a number of alternative reading modes have been introduced
to extend the computational power of finite state machines. The quest is two-
fold: on one hand one would like to obtain a class of machines which have all
the pleasant algorithmic and closure properties of finite state machines yet be
able to encompass a larger set of naturally occurring problems, on the other
hand, trying to map the limits of capabilities of machine models which cannot store information on a tape, thereby remembering only a constant-bounded
amount of information about the previously read input. Among these alternative automata models we find restarting automata [8], revolving automata [4],
jumping automata [9], automata with translucent letters [10] and one-way jumping automata [5]. In several cases it turns out that the extended machine has
the same power as classical deterministic finite automata (DFA), but even in
those cases there can be aspects such as reduced state complexity, which make it
worth studying the models in detail. One-way jumping automata was introduced
in 2016 and investigated since with respect to its accepting power and closure
properties [3,5], decidability questions [1], nondeterminism [2,6] and extensions
of the tape head movement to more powerful machine models [6]. Perhaps some-
what unexpectedly, it turns out that some decision problems related to one-way

M. Li (Ed.): FAW 2020, LNCS 12340, pp. 108–120, 2020.
https://doi.org/10.1007/978-3-030-59901-0_10

jumping DFA are difficult to answer, the most notable among them being equivalence and regularity.

Our contribution here stretches the inquiry to machines which can move in both directions. One of the aims is to try to complete a hierarchy of machines in this new input processing mode. Another is to try to find a storage-less machine model which can recognize arbitrary context-free languages, and which is simpler to use in proofs than push-down automata. We make progress with the first by describing certain basic properties of 2-way jumping finite automata and we mention a promising direction with regards to extensions which could encompass the context-free class, as well.

As is well-known since their introduction [11], two-way deterministic finite automata (2DFA) recognize the same class of languages as one-way automata. However, we also know [12] that one can exhibit a series of languages which can be accepted by 2DFA with exponentially fewer states than the minimal DFA. As we will see, in jumping mode, these automata are stricly stronger than in standard input reading mode. This is not surprising since already the language class accepted by one-way jumping automata strictly includes the regular class.

After recalling a few basic preliminaries, in Sect. 3 we define 2-way jumping DFA and investigate their basic properties including closure properties and we exhibit witness languages which prove that 2-way jumping DFA are strictly more powerful than one-way jumping DFA. In Sect. 4 we discuss the alternative ways of extending the one-way jumping movement to two directions and we show that they result in models with the same power. Finally, in Sect. 5, we address the problem of non-terminating computations in potential implementations of the new machines. Although here we only address this for one-way jumps, the approach naturally extends to two-way jumps. Some of the proofs reusing previously exhibited arguments as well as the detailed argument for the terminating version of one-way jumping machines have been omitted due to the space requirements.

2   Preliminaries

We recall notations of automata (see [7,13]). Σ generally stands for a finite alphabet, Σ^* for the set of all words over Σ and ϵ is the empty word. A *deterministic finite automaton* (DFA) M is a 5-tuple $(Q, \Sigma, \delta, s, F)$, where Q is the set of states, Σ is the input alphabet, $\delta : Q \times \Sigma \to Q$ is the transition relation, s is the initial state, and F is set of accepting states.

Here we do not require δ to be defined on the whole domain $Q \times \Sigma$, in fact, the additional power comes from the very fact that there are undefined transitions for certain state-letter pairs.

2.1   Tape Head Modes

Suppose M is a DFA.

Standard Mode

Suppose $q_1, q_2 \in Q$, $w \in \Sigma^*$ and s is the initial state. A configuration of M is a string in $Q \times \Sigma^*$. A transition from configuration $q_1 aw$ to configuration $q_2 w$, written as $q_1 aw \to q_2 w$, is possible when $q_2 = \delta(q_1, a)$. In the standard manner, we extend $\to$ to $\to^m$, where $m \geq 0$. Let $\to^+$ and $\to^*$ denote the transitive and the transitive-reflexive closure of $\to$, respectively.

A DFA with *standard mode* is a rewriting system $(M, \to)$ based on $\to^*$. The language accepted by $(M, \to)$ is $L(M, \to) = \{w \mid w \in \Sigma^*, sw \to^* f, f \in F\}$.

One-Way Jumping Mode

The *right one-way jumping relation* (denoted by $\circlearrowright_R$ here) between configurations from $Q\Sigma^*$, was defined in [5]. Let $x, y \in \Sigma^*$, $a \in \Sigma$ and $p, q \in Q$ such that $q = \delta(p, a)$. Then the right one-way jumping automaton M makes a jump from the configuration $pxay$ to the configuration qyx, symbolically written as $pxay \circlearrowright_R qyx$ if x belongs to $\{\Sigma \setminus \Sigma_p\}^*$ where $\Sigma_p = \{b \in \Sigma \mid \exists q \in Q \text{ s.t. } q \in \delta(p, b)\}$. In the standard manner, we extend $\circlearrowright_R$ to $\circlearrowright_R^m$, where $m \geq 0$. We denote by $\circlearrowright_R^*$ the transitive-reflexive closure of $\circlearrowright_R$. Intuitively, a machine in right one-way jumping mode will look for the closest letter to the right of its current position, for which it has a defined transition. This means that when the automaton is completely defined, then one-way jumping mode works the same way as the standard reading mode. While incomplete DFA have the same accepting power as complete ones in the classical case, in this new mode of tape head movement non-regular and even non-context-free languages can be accepted by incomplete finite state machines.

We define a DFA with *one-way jumping mode of tape head* to be a rewriting system $(M, \circlearrowright_R)$ based on $\circlearrowright_R^*$. The language accepted by $(M, \circlearrowright_R)$ is defined to be $L(M, \circlearrowright_R) = \{w \mid w \in \Sigma^*, sw \circlearrowright_R^* f, f \in F\}$.

3   Two-Way Jumping Mode

In this section we will define and investigate the properties of finite state machines which are allowed to jump in either direction when looking for the next input letter to read. We define the one way left jumping movement $\circlearrowright_L$, the counterpart of the previously defined $\circlearrowright_R$ as: (1) jump from right to left; (2) if the head can read the letter at current position, then it reads, otherwise move left; (3) when the head reaches the left end of the input, return to the right end.

An automaton in 2-way jumping mode will be able to jump to the left or right in each step depending on its transition function. Additionally, we will assume that the input tape is circular.

Definition 1 (2-way jumping deterministic finite automata)

1. *A 2-way jumping deterministic finite automata ($\circlearrowright_{RL} DFA$) is $(M, \circlearrowright_{RL}) = (Q, \Sigma, \delta, s, F)$, where Q, Σ, s, F are the same as in a usual DFA. The transition function is*

$$\delta : Q \times \{L, R\} \times \Sigma \to Q.$$

For convenience, we will denote state-direction pairs $p \in Q$, $d \in \{L, R\}$ by p^d, and the transition function is interpreted as $\delta(p^d, a) = q$ meaning that the *machine reads the first a (or some other letter if other transitions are defined for p, too) in the direction d, and changes its current state to q. In order for the execution to be deterministic, we need to restrict δ so that from each state there is only one direction in which the machine looks for the next letter to read. Formally, for each $p \in Q$ if there exists $d_1 \in \{L, R\}$, $a \in \Sigma$ and $q \in Q$ such that $\delta(p^{d_1}, a) = q$ then $\delta(p^{d_2}, b)$ is not defined for $d_2 \neq d_1$ for any letter $b \in \Sigma$.*

2. *Elements of $Q^{\{L,R\}}\Sigma^*$ are configurations of M. Let $\circlearrowleft_{RL} := \circlearrowleft_R \cup \circlearrowleft_L$ be the transition relation between configurations, given as*

$$p^{d_1} uav \circlearrowleft_{RL} q^{d_2} vu$$

when $u, v \in \Sigma^*$, $p, q \in Q$, $d_1, d_2 \in \{L, R\}$ and $\delta(p^{d_1}, a) = q$ such that one of the following is true:
(a) $d_1 = R$ and $u \in (\Sigma - \Sigma_p)^*$, or
(b) $d_1 = L$ and $v \in (\Sigma - \Sigma_p)^*$.
In other words $p^{d_1} uav \circlearrowleft_{RL} q^{d_2} vu$ if and only if $p^{d_1} uav \circlearrowleft_{d_1} q^{d_2} vu$.

3. *The language accepted by $(M, \circlearrowleft_{RL})$ is*

$$L(M, \circlearrowleft_{RL}) = \{w \in \Sigma^* \mid s^d w \circlearrowleft_{RL}^* f^{d_1} \text{ for some } f \in F \text{ and } d_1 \in \{L, R\}\}.$$

The usual input reading mode without jumping will be denoted by ($\rightarrow_{RL}$), the only distinction being that now we have two possible directions ($\rightarrow_R$ and $\rightarrow_L$) to read symbols from. So, for $p, q \in Q$, $w \in \Sigma^*$, $a \in \Sigma$:

1. If $p^R aw \circlearrowleft_R qw$, then we write $p^R aw \rightarrow_R qw$.
2. If $p^L wa \circlearrowleft_L qw$, then we write $p^L wa \rightarrow_L qw$.

Finally, $\rightarrow_{RL}$ is a reading step in either direction: $\rightarrow_{RL} := \rightarrow_R \cup \rightarrow_L$.

As we said in the abstract, two-way jumping automata are strictly more powerful than one-way jumping ones. The following examples demonstrate the increase in accepting power through two well-known non-regular, context-free languages, which can be accepted by the new model, but not by one-way jumping finite automata.

Example 1. Let M be a $\circlearrowleft_{RL}$DFA given by

$$M = (\{s, p, q\}, \{a, b\}, \delta, s, \{s\})$$

where δ is defined by $\delta(s^R, a) = p, \delta(s^R, b) = q, \delta(p^L, b) = s, \delta(p^L, a) = q$ (see Fig. 1). Then, $L(M, \circlearrowleft_{RL}) = \{a^n b^n \mid n \geq 0\}$. It is known that $\{a^n b^n \mid n \geq 0\}$ cannot be accepted by one-way jumping finite automata [5].

Example 2. Our second example further illustrates the power of 2-way jumping. The automaton on the right in Fig. 1, using an added end marker \$, accepts the language of balanced parenthesis pairs, i.e., the Dyck language with one pair of brackets. The example can be easily generalized to more pairs. This language

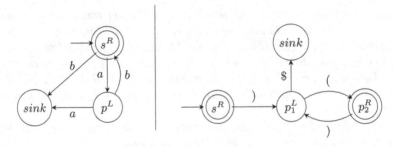

Fig. 1. Left: 2-way jumping DFA accepting $\{a^n b^n \mid n \geq 0\}$. Right: 2-way jumping DFA for the Dyck language with one bracket pair concatenated with an end marker \$.

is significant, because by the well-known Chomsky-Schützenberger characterization, each context-free language L can be expressed as $L = h(D \cap R)$, where D is a Dyck language, R is a regular language and h is a homomorphism. As we will see, the closure properties of 2-way jumping DFA do not allow construction of automata for arbitrary context-free languages based on this characterization, due to the class of languages accepted not being closed under either intersection with regular languages or applying homomorphisms. In fact, the class is incomparable with CFL witnessed by the languages $L_{a+ab} \in$ CFL$\setminus \circlearrowleft_{RL}$ **DFA** and $L_{abc} \in \circlearrowleft_{RL}$ **DFA** $\setminus$ CFL. However, a subsequent extension of the model incorporating nondeterminism and a mechanism for recognizing intersections with regular languages is one of our goals in follow-up research.

Theorem 1. *Let M be a $\circlearrowleft_{RL}$DFA and let $|\Sigma| = 1$, then $L(M)$ is regular.*

Proof. The argument is the same as in the case of one way jumping automata [5]. For any accepted unary input, the machine reads all of it without jumping, and reading towards the left results in the same remaining input as reading towards the right, so a classical DFA can accept the same language. □

Proposition 1. *Consider a $\circlearrowleft_{RL}$DFA $(Q, \Sigma, \delta, s, F)$ and let $p, q \in Q$, $w \in \Sigma^*$, $d_1, d_2 \in \{R, L\}$. If $p^{d_1}w \rightarrow^*_{RL} q^{d_2}$ holds, then there exist $w_1, w_2 \in \Sigma^*$ with $w_1 w_2 = w$ such that $p^{d_1} w_1 w' w_2 \rightarrow^*_{RL} q^{d_2} w'$ for any $w' \in \Sigma^*$.*

Proof. The argument is similar to the case of $\circlearrowleft_R$DFA (see [6], Proposition 1). We prove the statement by induction on the number of transitions. Let $p^{d_1}w \rightarrow^n_{RL} q^{d_2}$. We show that there exist w_1 and w_2 such that $w_1 w_2 = w$ and $p^{d_1} w_1 w' w_2 \rightarrow^n_{RL} q^{d_2} w'$ for any $w' \in \Sigma^*$. If $n = 0$, then $w = \epsilon$, $q = p$ and $d_1 = d_2$, and for any $w' \in \Sigma^*$, $p^{d_1} w' \rightarrow^0_{RL} q^{d_1} w'$ holds. Suppose that the statement holds for $n \leq k$. If $n = k + 1$, then there exist $r \in Q$, $d_3 \in \{R, L\}$ and $a \in \Sigma$ such that $p^{d_1}w \rightarrow^k_{RL} r^{d_3}a \rightarrow_{d_3} q^{d_2}$. This means that there exist $w_3, w_4 \in \Sigma^*$ with $w_3 a w_4 = w$ such that $p^{d_1} w_3 w_4 \rightarrow^k_{RL} r^{d_3}$. By the induction hypothesis, for any $u \in \Sigma^*$ we know that $p^{d_1} w_3 u w_4 \rightarrow^k_{RL} r^{d_3} u$ holds, so we can set $u = aw'$ or $w'a$ for some w', depending on d_3. We get that $p^{d_1} w_3 aw' w_4 \rightarrow^k_{RL} r^R aw' \rightarrow_R q^{d_2} w'$ or $pw_3 w' aw_4 \rightarrow^k_{RL} r^L w'a \rightarrow_L q^{d_2} w'$. In the first case, the statement holds with $w_1 = w_3 a$, $w_2 = w_4$, in the second case with $w_1 = w_3$, $w_2 = aw_4$. □

The language $L_{ab} = \{w \mid |w|_a = |w|_b\}$ was the first non-regular example of a $\circlearrowright_R$DFA language. It appears in several examples and proofs, hence we will refer to it everywhere as L_{ab}.

Corollary 1. *There is no* $\circlearrowright_{RL}$*DFA M that accepts* $L_{a+ab} = a^* \cup L_{ab}$.

Proof. Suppose that there exists $\circlearrowright_{RL}$ DFA $M = (Q, \Sigma, \delta, s, F)$ such that $L(M) = L_{a+ab}$. For all $n > |Q|$, there exists $p \in F$ and $d_1, d_2 \in \{R, L\}$ such that $s^{d_1} a^n \circlearrowright_{RL}^n p^{d_2}$. Each intermediary state between s and p needs to have a transition defined for a otherwise the machine would be stuck and reject. This means $s^{d_1} a^n \rightarrow_{RL}^n p^{d_2}$, which allows us to apply Proposition 1 with $w_1 = a^i$ and $w_2 = a^j$ for some $i + j = n$. Then, there are $q \in F$ and $d_3 \in \{R, L\}$ such that $s^{d_1} a^i b^n a^j \rightarrow_{RL}^n p^{d_2} b^n \rightarrow_{RL}^n q^{d_3}$. Using a simple pumping argument we get that there exist $m > 0$ and l such that $p^{d_2} b^l (b^m)^* \rightarrow_{RL}^* q^{d_3}$ holds, so $a^i b^l (b^m)^* a^j \subseteq L(M)$, contradicting $L(M) = L_{a+ab}$. □

Corollary 2. *There is no* $\circlearrowright_{RL}$*DFA that accepts* $L_{aab} = a^* L_{ab}$ *or* $L_{aab}^r = L_{ab} a^*$.

Proof. The proof is essentially the same as the proof of Corollary 1. □

Proposition 2. *Let M be a* $\circlearrowright_{RL}$*FA and suppose* $w_1, w_2, w_3, w_4 \in \Sigma^*$ *and* $d_1, d_2 \in \{R, L\}$. *If* $p^{d_1} w_1 \circlearrowright_{RL}^* q^{d_2} w_2$, *then there exist* w_3 *and* w_4 *such that* $w_3 w_2 w_4 \in perm(w_1)$ *and* $p^{d_1} w_3 w_2 w_4 \rightarrow_{RL}^* q^{d_2} w_2$ *holds, where* $perm(w_1)$ *is the set of permutations of* w_1.

Moreover, if $w \in L(M, \circlearrowright_{RL})$, *then there exist* $w' \in perm(w)$ *and* $d_1, d_2 \in \{R, L\}$ *such that* $w' \in L(M, \rightarrow_{RL})$ *and* $s^{d_1} w' \rightarrow_{RL}^* f^{d_2}$ *for some* $f \in F$.

Proof. The argument is similar to the case of one-way jumping machines [5,6]. If a jumping transition is performed, we can instead permute the input so that the letter read is next to the reading head and perform a standard transition resulting in a cyclic shift of the remaining input. □

Not surprisingly, the proposition above means that the languages accepted by $\circlearrowright_{RL}$DFA have a semilinear Parikh-image, as was the case for the one-way jumping machines. We can obtain a rather weak, but nevertheless potentially useful pumping lemma type result, too, along the same lines as in the case of one-way jumping machines, but we omit it here as it is not necessary for what follows.

Corollary 3. *There is no* $\circlearrowright_{RL}$*DFA M that accepts* $L_{a^n b^n c^n} = \{a^n b^n c^n : n \geq 0\}$.

Proof. Suppose that there exists $\circlearrowright_{RL}$DFA M such that $L(M, \circlearrowright_{RL}) = L_{a^n b^n c^n}$ and consider $a^n b^n c^n \in L_{a^n b^n c^n}$, where $n > |Q|$. For the first letter read by the machine, we have three cases, as follows.

(1) The head reads symbol b or c by $\circlearrowright_R$. If it reads a b by $\circlearrowright_R$, then by the definition of $\circlearrowright_R$, we have $s^R a^n b^n c^n \circlearrowright_R p_1^d b^{n-1} c^n a^n$ if and only if $s^R b^n c^n a^n \circlearrowright_R p_1^d b^{n-1} c^n a^n$ where $p_1 \in Q$ and $d \in \{R, L\}$. This means that M accepts $b^n c^n a^n$, a contradiction. If the first letter read was c, then the machine accepts $c^n a^n b^n$, again contradicting $L(M, \circlearrowright_{RL}) = L_{a^n b^n c^n}$.

(2) The head reads a or b by $\circlearrowleft_L$. Same as case (1).
(3) The head reads a or c by $\rightarrow_{RL}$. Before reading b, the automaton will read a's and c's, k times for some $k \geq 0$, that is:

$$s^{d_1} a^n b^n c^n \rightarrow^k_{RL} p_k^{d_2} a^{n-i} b^n c^{n-j}$$

where $i, j \in \mathbb{N}$ with $i + j = k$ and $d_1, d_2 \in \{R, L\}$. We have two cases:

(Case 1) if $k \leq |Q|$: we get $d_2 = R$ and $p_k^R a^{n-i} b^n c^{n-j} \circlearrowleft_R p_{k+1}^{d_3} b^{n-1} c^{n-j} a^{n-i}$ or $d_2 = L$ and $p_k^L a^{n-i} b^n c^{n-j} \circlearrowleft_L p_{k+1}^{d_3} c^{n-j} a^{n-i} b^{n-1}$ where $d_3 \in \{R, L\}$. We arrive at a contradiction by the same argument as in (1).
(Case 2) if $k > |Q|$: we get $s_1^d a^n b^n c^n \rightarrow^k_{RL} p_k^{d_2} a^{n-i} b^n c^{n-j}$ for some $d_1, d_2 \in \{L, R\}$. By a typical pumping argument, since $k > |Q|$, some state must repeat between s and p_k, so there are l, m with $l + m > 0$ such that each word from $a^{n-l} (a^l)^* b^n c^{n-m} (c^m)^*$ has a permutation in $L(M, \circlearrowleft_{RL})$, hence $L(M, \circlearrowleft_{RL}) \neq L_{a^n b^n c^n}$.

$\square$

Corollary 4. *There is no $\circlearrowleft_{RL} DFA$ M that accepts $\overline{L_{ab}} = \{a, b\}^* \setminus L_{ab}$.*

Proof. Suppose that there exists $\circlearrowleft_{RL}$DFA M such that $L(M, \circlearrowleft_{RL}) = \overline{L_{ab}}$. If M is completely defined, that is, if for each state $p \in Q$ there exists $d \in \{L, R\}$ and $q, r \in Q$ such that $\delta(p^d, a) = q$ and $\delta(p^d, b) = r$, then the machine is a classical 2DFA and hence the language it accepts is regular, whereas $\overline{L_{ab}}$ is not.

If M is not completely defined, then there exists some state p reachable from s for which $\delta(p^d, a)$ is not defined for either of $d \in \{L, R\}$ (the case of undefined b transition is symmetrical). Let a shortest path from s to p be labeled by a word w. For some permutation $w_1 w_2$ of w we have $s^{d_1} w_1 a^{|w|+1} w_2 \rightarrow^*_{RL} p^{d_2} a^{|w|+1}$ and M is stuck, rejecting the input even though $w_1 a^{|w_1 w_2|+1} w_2 \in \overline{L_{ab}}$ for all w_1, w_2.

$\square$

As a consequence of Corollaries 1–4 above about certain witness languages not being in the class accepted by $\circlearrowleft_{RL}$DFA, we can settle the status of the class with respect to closure under most of the common language operations.

Theorem 2. *The class $\circlearrowleft_{RL}$ **DFA** is not closed under the operations: (1) union, (2) union with regular language, (3) intersection, (4) intersection with regular language, (5) complementation, (6) concatenation, (7) right concatenation with regular language, (8) left concatenation with regular language, (9) homomorphism, (10) substitution with $\circlearrowleft_{RL}DFA$ languages, (11) permutation closure.*

Proof. Let $L_{ab} = \{w \in \{a, b\}^* \mid |w|_a = |w|_b\}$, and let $L_{a+ab}, L_{aab}, L_{a^n b^n c^n}, \overline{L_{ab}}$ be defined as in Corollary 1, 2, 3 and 4, respectively.

Since $a^* \cup L_{ab} = L_{a+ab} \notin \circlearrowleft_{RL}$ **DFA** by Corollary 1, (1) and (2) hold, because L_{ab} is a $\circlearrowleft_{RL}$ **DFA** language and a^* is regular. As L_{a+ab} is the permutation closure of the regular language $a^* \cup (ab)^*$, (11) also holds.

By Corollary 3 we have $L_{abc} \cap a^*b^*c^* = L_{a^nb^nc^n} \notin \circlearrowright_{RL}$ **DFA**, so we get (3) and (4), since L_{abc} is a $\circlearrowright_{RL}$ **DFA** language and $a^*b^*c^*$ is regular.

Corollary 2 says that $a^*L_{ab} = L_{aab}$ and $L_{ab}a^* = L_{aab}{}^r$ are not in $\circlearrowright_{RL}$ **DFA**, which means non-closure under concatenations, even with regular languages, so (6), (7) and (8) hold.

For (5), by Corollary 4 we have that $\overline{L_{ab}}$ does not belong to $\circlearrowright_{RL}$ **DFA**, so the class is not closed under complementation.

To see that applying a homomorphism h can lead outside the class, define $h : \{a,b,c\}^* \to \{a,b\}^*$ as $h(a) = a$, $h(b) = b$ and $h(c) = \epsilon$. Consider $L = \{w \in \{a,b\}^* \mid |w|_a = |w|_b\} \cup \{w \in \{a,c\}^* \mid |w|_a = |w|_c\}$, which is accepted by a $\circlearrowright_{RL}$ **DFA** with $\delta(s^R, b) = p^R$, $\delta(s^R, c) = q^R$ and $\delta(p^R, a) = s_b^R$, $\delta(q^R, a) = s_c^R$ and $\delta(s_b^R, b) = p^R$, $\delta(s_c^R, c) = q^R$, with initial and final state s and additional final states s_b, s_c. Then $h(L) = L_{a+ab} = \{w \in \{a,b\}^* \mid |w|_a = |w|_b\} \cup a^*$. Therefore (9) holds.

Finally, define the substitution from $\{a,b\}$ to $2^{\{a,b\}^*}$ as $\sigma(a) = a^*$ and $\sigma(b) = L_{ab}$. Both $\sigma(a)$ and $\sigma(b)$ can be accepted by $\circlearrowright_{RL}$ **DFA**, but $\sigma(ab) = \sigma(a)\sigma(b) = L_{aab}$ can not, therefore (10) holds. $\qquad\square$

Next, we look at closure under reversal, Kleene star and Kleene plus.

Theorem 3. *The class $\circlearrowright_{RL}$ **DFA** is closed under reversal.*

Proof. Let $(M, \circlearrowright_{RL}) = (Q, \Sigma, \delta, s^d, F)$ is $\circlearrowright_{RL}$DFA. Then, by simply reversing the direction of each state results in accepting the reverse language. The machine for it, $M^r = (Q', \Sigma, \delta', s^{d'}, F')$ will be defined by $Q' = \{q^{d_1} \mid q^{d_2} \in Q, d_1 \neq d_2\}$, $\delta'(p^{d_1}, a) = q^{d_2}$ if and only if $\delta(p^{d_3}, a) = q^{d_4}$ with $d_1 \neq d_3$ and $d_2 \neq d_4$, $d \neq e$, and $F' = \{q^{d_1} \mid q^{d_2} \in F, d_1 \neq d_2\}$.

$\qquad\square$

Theorem 4. *The class $\circlearrowright_{RL}$ **DFA** is not closed under Kleene star and Kleene plus.*

Proof. Consider $L_{a^nb^n} = \{a^nb^n : n \geq 0\}$, which is accepted by the $\circlearrowright_{RL}$DFA in Example 1. We show that there is no $\circlearrowright_{RL}$DFA which accepts $L_{a^nb^n}^*$ (the proof for $L_{a^nb^n}^+$ is the same). Suppose that there exists $\circlearrowright_{RL}$DFA M such that $L(M) = (L_{a^nb^n})^*$ and consider $a^mb^ma^mb^m \in (L_{a^nb^n})^*$, where $m > |Q|$. We have three cases: (1) head reads b by $\circlearrowright_R$; (2) head reads a by $\circlearrowright_L$; (3) head reads a number of a's and b's first by $\to_{RL}$ before jumping.

The argument in all three cases follows the proof of Corollary 3, and we omit them here due to space constraints. $\qquad\square$

4   Equivalence with Alternatively Defined $\circlearrowright_{RL}$DFA

The generalization of 2DFA to machines with $\circlearrowright_{RL}$ mode introduced in the previous section is not the only way one can extend the model. In this section we address this issue by defining another type of 2DFA with $\circlearrowright_{RL}$ mode and showing that the two definitions lead to machines with the same accepting power.

Previously the direction in which the automaton was looking for the next letter to read was determined by the current state. It is possible to define the model such that the direction of the next read is determined by the previous transition instead of the current state. The classical case is perhaps more naturally generalized to this new version, but the examples and arguments are simpler to describe in the previous model, hence our choice of the order of presentation. In this alternatively defined model:

- first, the head starts reading in the initial direction specified;
- later, the jump direction is determined by the value returned by δ in the previous transition.

Definition 2 (Alternative 2-way jumping finite automata). *An alternative $\circlearrowleft_{RL}$ DFA is a tuple $M = (Q, \Sigma, \delta, s, d, F)$ where Q, Σ, s, F are as usual and $d \in \{L, R\}$ is the initial direction. The transition function is*

$$\delta : Q \times \Sigma \to Q \times \{L, R\},$$

and it is interpreted as follows: if M reads 'a' when being in state p and $\delta(p, a) = (q, e)$, then in the next step the machine will apply a $\circlearrowleft_e$ transition, i.e., it will look for the next input letter in direction e. In the initial state s this direction is given by the initial direction d.

Example 3. Let M be a $\circlearrowleft_{RL}$ DFA given by

$$M = (\{s, p, q\}, \{a, b\}, \delta, s, R, \{s\})$$

where δ is given by $\delta(s, a) = (p, L)$, $\delta(s, b) = (q, L)$, $\delta(p, b) = (s, R)$, $\delta(p, a) = (q, R)$ and in the first step, the head jumps to the right (see Fig. 2). Then, $L(M, \circlearrowleft_{RL}) = \{a^n b^n \mid n \geq 0\}$.

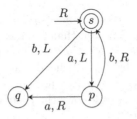

Fig. 2. Alternative 2-way jumping FA accepting $\{a^n b^n \mid n \geq 0\}$.

Theorem 5. *Every alternative $\circlearrowleft_{RL}$ DFA has an equivalent $\circlearrowleft_{RL}$ DFA and every $\circlearrowleft_{RL}$ DFA has an equivalent alternative $\circlearrowleft_{RL}$ DFA.*

Proof. Let $M = (Q, \Sigma, \delta, s, d, F)$ be an alternative $\circlearrowright_{RL}$DFA. For the purpose of the simulation below, we partition the set of states Q into Q_R, Q_L, Q_{RL}:

$$Q_{RL} = \{p \in Q \mid \exists\, q, r \in Q \text{ and } a, b \in \Sigma \text{ s.t. } \delta(q, a) = (p, R), \delta(r, b) = (p, L)\},$$

$$Q_R = \{p \in Q \setminus Q_{RL} \mid \exists q \in Q, a \in \Sigma \text{ s.t. } \delta(q, a) = (p, R)\},$$

$$Q_L = \{p \in Q \setminus (Q_{RL} \cup Q_R) \mid \exists q \in Q, a \in \Sigma \text{ s.t. } \delta(q, a) = (p, L)\}.$$

We add the initial state either to Q_L or to Q_R depending on the initial direction, so $Q = Q_R \cup Q_L \cup Q_{RL}$. We will construct a $\circlearrowright_{RL}$DFA $M' = (Q', \Sigma, \delta', s, F')$ such that $L(M) = L(M')$.

For each $p \in Q_{d_1}$, where $d_1 \in \{L, R\}$, we put p^{d_1} in Q' and we set $\delta'(p^{d_1}, a) = q^{d_2}$ if and only if $\delta(p, a) = (q, d_2)$. For the states $p \in Q_{RL}$ we make two copies p^L and p^R and place them in Q'. Then, for all $q \in Q$, $a \in \Sigma$ and $d \in \{L, R\}$, we set $\delta'(p^L, a) = \delta'(p^R, a) = q^d$ if and only if $\delta(p, a) = (q, d)$. Finally we set $F' = \{q^d \in Q' \mid q \in F\}$ (see Fig. 3).

For each transition in M from p to q labeled by (a, d) there is a corresponding transition in M' from p^L and/or p^R to q^d labeled by a and vice versa, so the accepted inputs are the same.

The reverse direction is straightforward. Assume that $M' = (Q', \Sigma, \delta', s^d, F')$ is a $\circlearrowright_{RL}$DFA as defined initially. Construct the alternative $\circlearrowright_{RL}$DFA $M' = (Q, \Sigma, \delta, s, d, F)$ by setting $Q' = Q$, $F = F'$ and $\delta(p, a) = (q, d)$ if and only if $\delta'(p^e, a) = q^d$ for some $p, q \in Q$, $a \in \Sigma$ and $e, d \in \{L, R\}$. $\qquad\square$

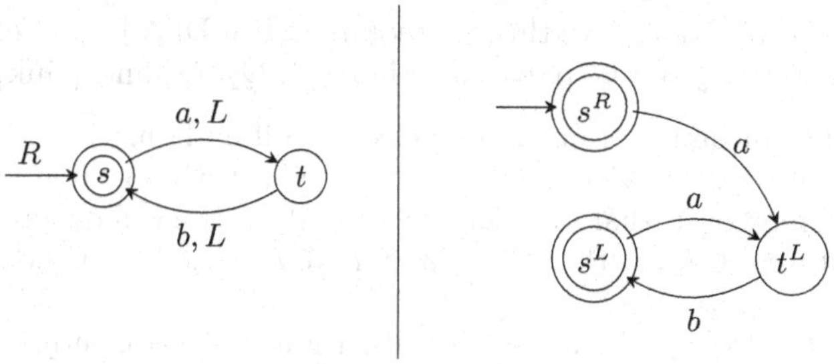

Fig. 3. Left: Alternative $\circlearrowright_{RL}$DFA for $L_{ab} = \{w \in \{a, b\}^* \mid |w|_a = |w|_b\}$. Right: $\circlearrowright_{RL}$DFA simulating M, accepting L_{ab}.

5   Avoiding Infinite Loops

Finite automata are the simplest model of physical computing machines and are implemented in many electromechanical devices such as vending machines and elevators, and software systems like text editors and compilers. Finite automata and other deterministic automata always stop after a finite number of steps, on the other hand, an automaton $(M, \circlearrowright_R)$ with one-way jumping mode does not always terminate, as it may be caught in an infinite loop when there are still letters on the input tape, but the current state has no transition defined for any of them. A basic expectation of a non-Turing complete machine model from an implementation point of view is that the machines stop on any input and as such falling into an infinite loop is a defect for a physical model of a computing machine solving decision problems.

To remedy the infinite loop issue, we introduce a variant of one-way jumping mode that guarantees terminating computations, while keeping the accepting power identical to the machines of type $(M, \circlearrowright_R)$. In the new mode, we will place a marker at the end of the input word. This marker will not be erased after reading, thereby always serving as an anchor point. Suppose $\$$ is a symbol such that $\$ \notin \Sigma$. The language accepted by an automaton $(M, \circlearrowright^{ap})$ is defined to be

$$L(M, \circlearrowright^{ap}) = \{w \in \Sigma^* \mid sw\$ \circlearrowright^* f\$, f \in F, \$ \notin \Sigma_f\}$$

where $\Sigma_f = \{b \in \Sigma \mid \delta(f, b) = q \text{ for some } q \in Q\}$. The symbol $\$$ plays a role as an *anchoring point* of the input word which is formatted as a circuit in one-way jumping mode. One-way jumping mode with anchoring point provides a terminating property similar to $(M, \rightarrow)$. Introducing an end marker which can be read multiple times changes the accepting power of DFA in $\circlearrowright_R$ mode. Take, for instance, the language $L = \{cw \mid |w|_a = |w|_b\}$. In [5] it was shown that there is no $\circlearrowright_R$DFA which accepts L^*, but making use of the $\$$ end marker a DFA can accept L^* in $\circlearrowright_R^{ap}$ mode (see Fig. 4).

To keep the accepting power the same as $\circlearrowright_R$DFA, we will restrict the use of the end marker strictly to checking whether there are no input letters readable in the current state. To achieve this, we require that a DFA in $\circlearrowright_R^{ap}$ mode is such that its set of states Q can be partitioned into Q_1, Q_2, Q_3 and $\{sink\}$ such that:

1. A state q has incoming transition labeled by $\$$ if and only if $q \in Q_2 \cup Q_3$.
2. For each state $q \in Q_i$ with $2 \le i \le 3$, there is exactly one state p such that $\delta(p, \$) = q$; moreover, that state satisfies $p \in Q_{i-1}$. We will denote $\text{pre}(q) = p$.
3. For each state $q \in Q_i$ with $2 \le i \le 3$ and all $a \in \Sigma$, there is no state $p \in Q$ with $\delta(p, a) = q$.
4. For all $q \in Q_1$ and $a \in \Sigma$, we require either $\delta(q, a) \in Q_1$ or $\delta(q, a)$ undefined. Moreover, $\delta(q, \$)$ is defined for all $q \in Q_1$.
5. For all $q \in Q_2$ and $a \in \Sigma$, we require $\delta(q, a) = \delta(\text{pre}(q), a)$.
6. For all $q \in Q_3$ and $a \in \Sigma$, we require $\delta(q, a) = \text{sink}$.
7. For all $a \in \Sigma$, we require $\delta(\text{sink}, a) = \text{sink}$ and $\delta(sink, \$)$ undefined.
8. The final condition is that $F \subseteq Q_3$.

With the above restriction in place we can show that for each $\circlearrowright_R$DFA there exists a DFA in $\circlearrowright_R^{ap}$ mode which accepts the same language, and vice versa, and the counterpart machines can be effectively constructed. The new machines operating in $\circlearrowright_R^{ap}$ always read the whole input and stop when there is only the end marker $\$$ on the tape and the current state has no transition defined for $\$$. For an example of how the anchor point construction, see Fig. 5.

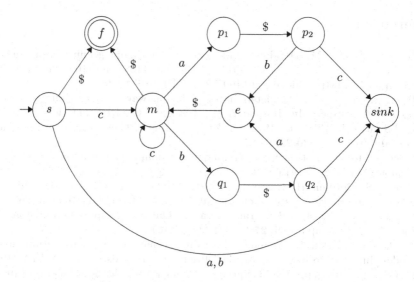

Fig. 4. $\circlearrowleft_R$ with non-erasable end marker \$ accepting L^*, where $L = \{cw \mid |w|_a = |w|_b\}$.

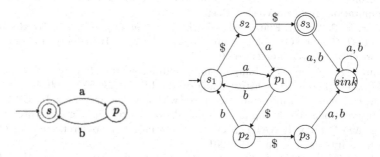

Fig. 5. Left: $\circlearrowleft_R$DFA for $L_{ab} = \{w \in \{a,b\}^* \mid |w|_a = |w|_b\}$. Right: $\circlearrowleft_R^{ap}$ DFA with anchoring point for L_{ab}. States with subscript i are in Q_i.

Theorem 6

(1) For any DFA M there exists a DFA M' such that $L(M, \circlearrowleft_R) = L(M', \circlearrowleft_R^{ap})$.
(2) For any DFA M' there exists a DFA M such that $L(M', \circlearrowleft_R^{ap}) = L(M, \circlearrowleft_R)$.

Proof. Simulating $\circlearrowleft_R$DFA with the new type of machines is trivial, while the end marker makes it possible to check for an otherwise empty tape. As for the other direction, the restrictions are tailored for the simulation of the new mode by $\circlearrowleft_R$DFA, where the Q_1 states of M' will be the states of M and states in Q_2, Q_3 will be discarded. Due to space constraints, the full argument is omitted. □

References

1. Beier, S., Holzer, M.: Decidability of right one-way jumping finite automata. In: Hoshi, M., Seki, S. (eds.) DLT 2018. LNCS, vol. 11088, pp. 109–120. Springer, Cham (2018). https://doi.org/10.1007/978-3-319-98654-8_9
2. Beier, S., Holzer, M.: Nondeterministic right one-way jumping finite automata (extended abstract). In: Hospodár, M., Jirásková, G., Konstantinidis, S. (eds.) DCFS 2019. LNCS, vol. 11612, pp. 74–85. Springer, Cham (2019). https://doi.org/10.1007/978-3-030-23247-4_5
3. Beier, S., Holzer, M.: Properties of right one-way jumping finite automata. Theoret. Comput. Sci. **798**, 78–94 (2019)
4. Bensch, S., Bordihn, H., Holzer, M., Kutrib, M.: On input-revolving deterministic and nondeterministic finite automata. Inf. Comput. **207**(11), 1140–1155 (2009)
5. Chigahara, H., Fazekas, S.Z., Yamamura, A.: One-way jumping finite automata. Int. J. Found. Comput. Sci. **27**(3), 391–405 (2016)
6. Fazekas, S.Z., Hoshi, K., Yamamura, A.: Enhancement of automata with jumping modes. In: Castillo-Ramirez, A., de Oliveira, P.P.B. (eds.) AUTOMATA 2019. LNCS, vol. 11525, pp. 62–76. Springer, Cham (2019). https://doi.org/10.1007/978-3-030-20981-0_5
7. Hopcroft, J.E., Ullman, J.D.: Introduction to Automata Theory, Languages and Computation. Addison-Wesley, Boston (1979)
8. Jančar, P., Mráz, F., Plátek, M., Vogel, J.: Restarting automata. In: Reichel, H. (ed.) FCT 1995. LNCS, vol. 965, pp. 283–292. Springer, Heidelberg (1995). https://doi.org/10.1007/3-540-60249-6_60
9. Meduna, A., Zemek, P.: Jumping finite automata. Int. J. Found. Comput. Sci. **23**(7), 1555–1578 (2012). https://doi.org/10.1142/S0129054112500244
10. Nagy, B., Otto, F.: Finite-state acceptors with translucent letters. In: BILC 2011–1st International Workshop on AI Methods for Interdisciplinary Research in Language and Biology, ICAART 2011 – 3rd International Conference on Agents and Artificial Intelligence, pp. 3–13 (2011)
11. Rabin, M.O., Scott, D.: Finite automata and their decision problems. IBM J. Res. Dev. **3**(2), 114–125 (1959)
12. Shepherdson, J.C.: The reduction of two-way automata to one-way automata. IBM J. Res. Dev. **3**(2), 198–200 (1959)
13. Sipser, M.: Introduction to the Theory of Computation, 2nd edn. Course Technology, Boston (2006)

A Loopless Algorithm for Generating (k, m)-ary Trees in Gray-Code Order

Yu-Hsuan Chang[1], Ro-Yu Wu[2], Cheng-Kuan Lin[3], and Jou-Ming Chang[1(✉)]

[1] Institute of Information and Decision Sciences, National Taipei University
of Business, Taipei, Taiwan
{10766004,spade}@ntub.edu.tw
[2] Department of Industrial Management, Lunghwa University of Science
and Technology, Taoyuan, Taiwan
eric@mail.lhu.edu.tw
[3] College of Mathematics and Computer Science,
Fuzhou University, Fuzhou 350108, China
cklin@fzu.edu.cn

Abstract. Introduced by Du and Liu in 2007, a (k, m)-ary tree is a generalization of k-ary tree such that the nodes have degree k on even levels and the nodes have degree 0 or m on odd levels. Especially, every node with the degree m on odd levels is called a crucial node. In a (k, m)-ary tree of order n, there are exactly n crucial nodes. A loopless algorithm is an algorithm for generating combinatorial objects using only assignment and comparison statements and does not include loop structure or recursion. In this paper, we propose a loopless algorithm to generate (k, m)-ary trees of order n in Gray-code order using Z-sequence suggested by Zaks in 1980. The required memory space of our algorithm is $2n + \mathcal{O}(1)$. Moreover, we analyze both amortized costs of assignment statements and comparison statements, which are no more than $3 + \frac{3}{km}$ and $2.5 + \frac{2}{km}$, respectively.

Keywords: Loopless algorithms · (k, m)-ary trees · Amortized analysis · Gray-codes · Zaks' sequences

1 Introduction

For quite some time, exhaustive algorithms for generating combinatorial objects (or objects for short) in multiple classes have been interested in computer science due to the requirements of practical applications, such as counterexample searching, combinatorial group testing, and algorithm performance analyzing. In an exhaustive algorithm of generating certain objects, the integer or alphabet sequence for representing an object is called a *codeword*. Encoding objects into variously specific codewords has been investigated widely, such as P-sequences [13], RD-sequences [17,18], Z-sequences [9,19–22], left-weight sequences [5,7,12,15,16], and left-child sequences [6,15]. And, the listing of these

© Springer Nature Switzerland AG 2020
M. Li (Ed.): FAW 2020, LNCS 12340, pp. 121–132, 2020.
https://doi.org/10.1007/978-3-030-59901-0_11

codewords is constituted in a particular order. For instance, the typical orderings to enumerate objects are B-order [1,2,8], lexicographical order [6,22], and gray-code order [13,16]. Hereafter, for convenience, we use the term object and its corresponding codeword (or sequence) interchangeably.

To efficiently enumerating objects, the main consideration is attempting to generate the corresponding sequence in constant time. A well-known ordering that can change consecutive sequences in constant time is the so-called *Gray-code*, which encodes objects into sequences so that the change between two consecutive sequences is restricted to only one position. Ehrlich [4] proposed a notion called the loopless algorithm for generating objects. In this type of algorithms, after an object is initialized, it can only execute assignment statements and "if-then-else" statements, and besides, it does not include loop structure or recursion. In loopless algorithms, as the calculation cost of generating the next object from the previous one only takes constant time, Gray-code order is usually concerned with.

The first loopless algorithm for generating objects in Gray-code was offered by Williamson [14]. For an excellent survey for listing Gray-codes of combinatorial objects, the readers can refer to Savage [10]. For binary trees, Vajnovszki proposed the loopless generations to enumerate their Gray-codes using the left-weight sequences [12] and P-sequences [13], respectively, where both sequences were introduced by Pallo [7,8]. For k-ary trees, the loopless Gray-code generation of left-weight sequences was presented by Korsh and LaFollette [5]. Then, Roelants van Baronaigien [9] and Xiang et al. [20,21] proposed the loopless algorithm for listing Gray-codes of k-ary trees using Z-sequences introduced by Zaks [22]. Later on, using the same Z-sequences, Wu et al. [19] proposed an improved loopless Gray-code generation of k-ary trees, which speeds up the generation and uses less memory space.

In this paper, we study the problem of generating (k, m)-ary trees (formally defined later in Sect. 2) in Gray-code order using Z-sequences. Du and Liu [3] first introduced the family of (k, m)-ary tree as a generalization of k-ary trees such that the degree of each internal node is determined based on the level of the node resided. In particular, a (k, m)-ary tree is said to be of order n if it has exactly n internal nodes on odd levels. Recently, Amani and Nowzari-Dalini [1] presented a recursive algorithm for generating all (k, m)-ary trees of order n represented by Z-sequences in B-order. In particular, each sequence in this algorithm is generated in a constant amortized time and $\mathcal{O}(n)$ time complexity in the worst case. Moreover, based on this ordering, they also proposed ranking and unranking algorithms. Shortly afterward, Chang et al. [2] provided new improvements for such ranking and unranking. Nevertheless, the loopless generation for (k, m)-ary trees of order n has not been developed yet until now. Thus, this inspires us to propose a loopless algorithm for generating (k, m)-ary trees of order n in Gray-code order using Z-sequences. As a consequence, the loopless generation in our algorithm requires $2n + \mathcal{O}(1)$ memory space. Moreover, for each generation, there are 3 (resp. 6) assignments needed in the best case (resp. worst case), and there are 2 (resp. 5) comparisons needed in the best case (resp. worst case) Additionally, we dissect the amortized time of assignment and comparison statements in each generation, which are bounded with no more than the constants $3 + \frac{3}{km}$ and $2.5 + \frac{2}{km}$, respectively.

2   Preliminaries

A *tree* is a connected acyclic graph. A node in a tree without children is called an *external node* (or *leaf*), otherwise, it is an *internal node*. A tree is *rooted* if one of its internal node, called the *root* and denoted by r, is distinguished from the others. A node v in a tree T is said to be *on the level* ℓ if the unique path from r to v has length ℓ, and the root r is on the level 0. A rooted tree is *ordered* if the children of each internal node have a designated order. A *k-regular tree* is a rooted tree with exactly k children on each internal node. A *k-ary tree* is an ordered k-regular tree (i.e., every internal node has exactly k-ordered children). A *k-ary tree of order* n is a k-ary tree that has exactly n internal nodes. Let $\mathcal{T}_k(n)$ denote the set of k-ary trees of order n. The number of k-ary trees of order n can be calculated by the generalized Catalan number [11], i.e., $|\mathcal{T}_k(n)| = C_k(n) = \frac{1}{kn+1}\binom{kn+1}{n}$.

A (k, m)-ary tree is a generalization of k-ary tree such that the degree of each internal node is determined by the level of the node resided. The precise definition is as follows.

Definition 1 (see [3]). For $k, m \geqslant 1$ and $n \geqslant 0$, a (k, m)-*ary tree of order* n is an ordered tree, such that

1) All nodes on even levels have degree k (where the root is on level 0).
2) All nodes on odd levels have degree m or 0, and there are exactly n nodes of degree m on all odd levels.

Particularly, the n internal nodes on odd levels are called *crucial nodes*. Let $\mathcal{T}_{k,m}(n)$ denote the set of (k, m)-ary trees with n crucial nodes. Du and Liu [3] also defined (k, m)-*Catalan number of order* n as $C_{k,m}(n)$, and proved that (k, m)-ary trees can be counted by (k, m)-Catalan numbers as follows.

Theorem 1 (see [3]). *The number of* (k, m)-*ary trees of order* n *can be computed by*

$$|\mathcal{T}_{k,m}(n)| = C_{k,m}(n) = \frac{1}{mn+1}\binom{(mn+1)k}{n}. \tag{1}$$

In 1980, Zaks [22] suggested an encoding scheme, called *Z-sequences*, for representing k-ary trees. For a k-ary tree T of order n, the encoding scheme labels the visited order of each internal node of T by traversing both internal nodes and leaves of T in preorder (i.e., an ordering of nodes visited from the root and then recursively the subtrees from left to right). Similarly, for a (k, m)-ary tree T of order n, the Z-sequence of T is encoded by traveling all nodes (including leaves and crucial nodes) on odd levels in preorder such that the visited order of each crucial node $i \in T$ is denoted by z_i, and the resulting Z-sequence of T is denoted by $z(T) = (z_1, z_2, \ldots, z_n)$. For example, the number of $(3, 2)$-ary trees of order 4 is equal to $\frac{1}{9}\binom{27}{4} = 1950$, and Fig. 1 shows three of them and their corresponding Z-sequences.

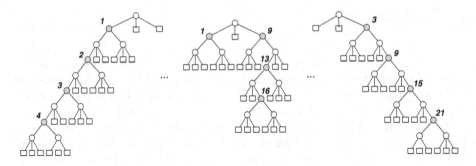

Fig. 1. The $(3, 2)$-ary trees of order 4 and their corresponding Z-sequences.

Due to the huge amount of (k, m)-ary trees of order n (even if all k, m, n are small integers), a systematical way to describe all Z-sequences of those trees is customarily the use of a *recursion tree*. In a recursion tree, each node on level i (except for the root which is on level 0) is associated with a label z_i so that the collection of labels along each path from a node on level 1 to a leaf (i.e., a node on level n) represents a Z-sequence corresponding to a (k, m)-ary tree of order n. Figure 2 exhibits the recursion tree for $(3, 2)$-ary trees of order 2. In this figure, the appearance of all children with a common parent is called a *fragment* on a level. Since the labels from left to right in each fragment are in increasing order in the recursion tree, the list of Z-sequences results in lexicographical order.

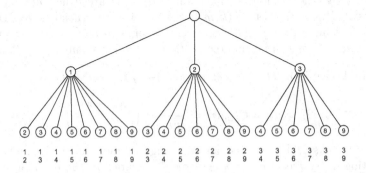

Fig. 2. The recursion tree for representing Z-sequences of $(3, 2)$-ary trees of order 2 in lexicographical order.

3   A Loopless Algorithm

For a tree $T \in \mathcal{T}_{k,m}(n)$, the *right-arm* of T is the path from the root to its rightmost leaf. In this section, we propose a loopless generation of (k,m)-ary trees in Gray-code order using Z-sequences. Before this, we need the following structure property of (k,m)-ary trees of order n.

Lemma 1. *Let T be a (k,m)-tree of order n and $z(T) = (z_1, z_2, \ldots, z_n)$ be the Z-sequence of T. Then, $z_{i-1} + 1 \leqslant z_i \leqslant km(i-1) + k$ for $i = 1, 2, \ldots, n$, where $z_0 = 0$.*

Proof. The proof is by induction on n. For $n = 1$, it is clear from Definition 1 that we have $1 \leqslant z_1 \leqslant k$. Suppose that $n \geqslant 2$ and the lemma holds for $n-1$, i.e., $z_{i-1} + 1 \leqslant z_i \leqslant km(i-1) + k$ for $i = 1, 2, \ldots, n-1$. We now consider Z-sequences for (k,m)-trees of order n, and the proof only needs to determine the range for the encoding of z_n. For each tree $T \in \mathcal{T}_{k,m}(n)$, since labels of crucial nodes are encoded by traversing odd levels of T in preorder, there are two cases as follows.

   Case 1: The node $n-1$ is contained in the right-arm of T. In this case, nodes $n-1$ and n are in different odd levels such that n is a grandchild of $n-1$. By the definition of encoding scheme, the next crucial node of $n-1$ is labeled by z_n such that $z_{n-1} + 1 \leqslant z_n \leqslant z_{n-1} + km$, where km denotes the number of grandchildren of $n-1$. By induction hypothesis, we have $z_n \leqslant z_{n-1} + km \leqslant (km(n-2) + k) + km = km(n-1) + k$. In particular, the node n is contained in the right-arm of T if $z_n = km(n-1) + k$.

   Case 2: The node $n-1$ is not contained in the right-arm of T. If $n-1$ is the grand parent of n, an argument similar to Case 1 shows that $z_{n-1} + 1 \leqslant z_n \leqslant z_{n-1} + km$. Otherwise, we have $z_{n-1} + km + 1 \leqslant z_n \leqslant km(n-1) + k$, where the upper range is obtained from Case 1. Then, combining the two situations, we determine that $z_{n-1} + 1 \leqslant z_n \leqslant km(n-1) + k$.   □

   Recall that a Gray-code generation is to enumerate sequences in a family of combinatorial objects such that a sequence differs from only one bit position with its predecessor. By Lemma 1, we note that for each fragment in the level i of a recursion tree, it always contains the largest label $km(i-1) + k$ and the second-largest label $km(i-1) + k - 1$ (e.g., see Fig. 2, we have labels 3 and 2 in the fragment on level 1, and labels 9 and 8 in the fragments on level 2). Thus, to construct a recursion tree in a Gray-code order, we can naturally readjust the order of nodes in each fragment such that the two extreme labels are regarded as two ends of a fragment. For example, Fig 3 exhibits a recursion tree for representing Z-sequences of $(3,2)$-ary trees of order 2 in Gray-code order.

   Further, two fragments on level i with particular ends in a recursion tree are called the *up-fragment* and the *down-fragment*, respectively, if they have the following permutations:

$$km(i-1) + k, z_{i-1} + 1, z_{i-1} + 2, \ldots, km(i-1) + k - 2, km(i-1) + k - 1$$

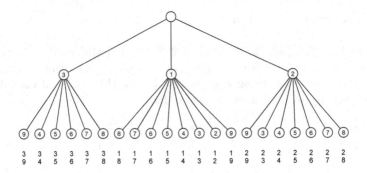

Fig. 3. A recursion tree for representing Z-sequences of (3,2)-ary trees of order 2 in Gray-code order.

and

$$km(i-1) + k - 1, km(i-1) + k - 2, \ldots, z_{i-1} + 2, z_{i-1} + 1, km(i-1) + k.$$

Note that the term z_{i-1} in the above two permutations means the label of the common parent for all children in the fragment. In particular, we set $z_0 = 0$ for the root of the recursion tree. According to this, we can readjust the permutation of labels in a fragment by using only the two kinds of fragments such that two consecutive fragments share a common label in their adjacent nodes. Consequently, all fragments on each level of the recursion tree are laid out in increasing order and decreasing order alternately. Particularly, we attempt to choose an up-fragment as the leftmost fragment on each level. Thus, we set the Z-sequence of the leftmost path in the recursion tree to be $(z_1, z_2, \ldots, z_n)$ where $z_i = km(i-1)+k$ for $1 \leqslant i \leqslant n$. In order to see the structure of the recursion tree more thoroughly, an extension of Fig. 3 is shown in Fig. 4 (here we omit the root in the drawing).

Lemma 2. *Two consecutive Z-sequences in the recursion tree differ in exactly one bit position.*

Proof. Suppose that the recursion tree has totally n levels (which does not include the root). Let u and v be two leaves adjacent to each other. By the arrangement of the recursion tree, there are two cases as follows.

Case 1: u and v have different labels. We can deduce that these two nodes have the same parent, denoted by w. Apparently, the labels collected from root to w for u and v are the same. This shows that the two consecutive sequences differ only at nth position.

Case 2: u and v have the same label. Let u' and v' be the lowest ancestors of u and v, respectively, such that their labels are different. Suppose that u' and v' are resided on level k where $1 \leqslant k < n$. Then, by Case 1, the two subsequences from the node on level 1 to u' and to v' differ only at kth position. Since the two Z-sequences, one is for u and the other is for v, are extended from these two

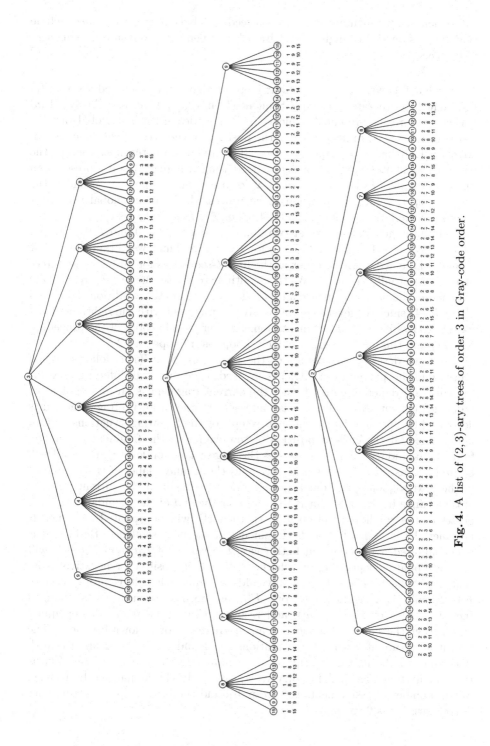

Fig. 4. A list of $(2, 3)$-ary trees of order 3 in Gray-code order.

subsequences by appending the same succeeding labels, respectively, we conclude that there is one-bit changing at position k between the two resulting sequences of length n. □

In what follows, we introduce our loopless algorithm. A procedure NextZ() is designed to generate the next Z-Sequence from the previous one. Array $z[1..n]$ is used to store the present Z-sequence in Gray-codes order. The global variable i (where $1 \leqslant i \leqslant n$) indicates where the position of the array z will be changed. Array $F[1..n]$ has two purposes in use. The first is to determine whether the current fragment $F[i]$ is increasing or decreasing, where $F[i] > 0$ represents an up-fragment and $F[i] < 0$ represents a down-fragment. And, the second is the use to keep tracking the position to be processed. Also, the global variable ℓ stands for the largest label in the level i (i.e., $\ell = km(i-1) + k$).

Before the generation, we set two dummy values $z[0] = 0$ and $F[0] = 1$, and let $z[j] = km(j-1) + k$ and $F[j] = j$ for $1 \leqslant j \leqslant n$. Thus, we obtain the largest sequence in lexicographical order as the first Z-sequences for all (k, m)-ary trees of order n. Initially, set $i = n$. We then continue to invoke the procedure NextZ() to generate the next Z-sequence until the last one meets the conditions $i = 1$ and $z[1] = k-1$ (since the level 1 contains only one fragment with the rightmost label $k-1$). To facilitate the implementation, we offer pseudo codes of the procedure NextZ() and the main program namely Loopless in Appendix A.

Now, we describe the details of the procedure NextZ() as follows. When every time the procedure NextZ() is invoked, there are two different rules for changing $z[i]$ depending on whether the current fragment is an up-fragment or a down-fragment. Normally, $z[i]$ is increased by one in an up-fragment and is decreased by one in a down-fragment except for the following situations. For an up-fragment, if the current position of the change in the fragment is starting from the boundary (i.e., $z[i]$ possesses the largest label on level i), then $z[i]$ will be updated as $z[i-1] + 1$ (i.e., it becomes the smallest label of the fragment) in the next sequence. In contrast, for a down-fragment, if the current position of the change in the fragment is to encounter the end of the boundary (i.e., $z[i]$ possesses the smallest label on level i), then $z[i]$ will be updated as ℓ (i.e., it becomes the largest label of the fragment) in the next sequence. Besides, if it is encountered a boundary in both cases, the values of $F[i]$ and $F[i-1]$ will also be updated depending on whether $F[i-1]$ is positive or negative. Note that the value of $F[i-1]$ probably adds a minus when it is propagated to $F[i]$. Finally, it designates $F[n]$ or $-F[n]$ as the next position to be changed (i.e., $i = F[n]$ or $i = -F[n]$), and recovers $F[n]$ to be n or $-n$ depending on whether the current fragment is an up-fragment or a down-fragment. The algorithm terminates when the conditions $i = 1$ and $z[1] = k - 1$ are attained. According to the above description, we summarize the procedure NextZ() as the flowchart in Fig. 5. Additionally, we can easily check that the best (resp. worst) number of assignments is 3 (resp. 6), and the best (resp. worst) number of comparisons is 2 (resp. 5).

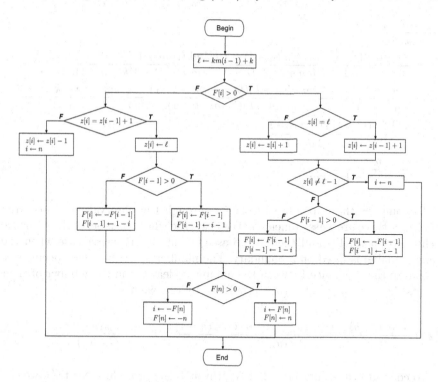

Fig. 5. The flowchart of the procedure NextZ().

Since the above procedure is inspired from Williamson's algorithm, the correctness of the loopless algorithms directly follows. Furthermore, the listing of all sequences is arranged in Gray-codes order which is certified by Lemma 2. Accordingly, we can prove the following main theorem.

Theorem 2. *Generating Z-sequences of (k,m)-ary trees with n internal nodes in Gray-code can be done in $\mathcal{O}(|T_{k,m}(n)|)$ time using $2n + \mathcal{O}(1)$ memory space. In addition, it requires amortized cost with at most $3 + \frac{3}{km}$ assignments and $2.5 + \frac{2}{km}$ comparisons, respectively, for each generation.*

Proof. The time and space complexities can easily be deduced from the loopless algorithm and the procedure NextZ(). Let the average number of assignments and the average number of comparisons be N_a and N_c, respectively. Before the analyses of N_a and N_c, we need the following bound of factor.

$$\frac{C_{k,m}(n-1)}{C_{k,m}(n)}$$

$$= \frac{mn+1}{m(n-1)+1} \times \frac{((m(n-1)+1)k)!\, n!\, ((mn+1)k-n)!}{((mn+1)k)!\, (n-1)!\, ((m(n-1)+1)k-n+1)!}$$

$$= \frac{mn+1}{mn-m+1} \times \frac{n(kmn-km+k)(kmn-km+k-1)\cdots(kmn-km+k-n+2)}{(kmn+k)(kmn+k-1)\cdots(kmn+k-n+1)}$$

$$= \frac{n\prod_{i=1}^{n-2}(m(n-1)+1)k-i}{\prod_{i=-1}^{n-1}(mn+1)k-i}$$

$$\leqslant \frac{n}{kmn+k-1} \leq \frac{1}{km}. \tag{2}$$

To compute the upper bound of N_a, through the observation of occurred situations, there are 3 assignments in the best case when the label of the sequence is changed on level n and there are 6 assignments in the worst case when the change of label is faced on a boundary. The number of the best case occurrence is equal to the difference between the number of leaves and the number of their parents (i.e., $C_{k,m}(n) - C_{k,m}(n-1)$). Thus, by (2), we have

$$N_a = \frac{3\cdot(C_{k,m}(n)-C_{k,m}(n-1))+6\cdot C_{k,m}(n-1)}{C_{k,m}(n)} = 3+3\frac{C_{k,m}(n-1)}{C_{k,m}(n)} \leqslant 3+\frac{3}{km}.$$

To compute the upper bound of N_c, through the procedure NextZ(), we can see that the best case in an up-fragment has 3 comparisons and in a down-fragment has 2 comparisons. And the worst case in an up-fragment has 5 comparisons and in a down-fragment has 4 comparisons. Similarly, the best case (respectively, the worst case) occurs in the situations when the changed position is on level n (respectively, the change of label is faced on a boundary). In either case, since up-fragments and down-fragments alternately appear on each level of the recursion tree, each type of fragments occupies for the half of the arrangement. Thus, by (2), we can show that

$$N_c = \frac{2.5\cdot(C_{k,m}(n)-C_{k,m}(n-1))+4.5\cdot C_{k,m}(n-1)}{C_{k,m}(n)} = 2.5+2\frac{C_{k,m}(n-1)}{C_{k,m}(n)} \leqslant 2.5+\frac{2}{km}.$$

$\square$

4   Concluding Remarks

In this paper, we developed a loopless algorithm for generating (k,m)-ary trees of order n encoded by Z-sequences in Gary-codes order. Each generation for Z-sequences can be done in constant time, and the memory space requires $2n + \mathcal{O}(1)$. In fact, the average number of assignments and the average number of comparisons for each tree in our loopless algorithm are at most $3 + \frac{3}{km}$ and $2.5 + \frac{2}{km}$, respectively.

Acknowledgments. This research was partially supported by MOST grants 108-2221-M-262-001 (R.-Y. Wu) and 107-2221-E-141-001-MY3 (J.-M. Chang) from the Ministry of Science and Technology, Taiwan.

Appendix A: the pseudo codes of the procedure NextZ() and the algorithm Loopless

Procedure NextZ( )

```
 1 begin
 2 |   ℓ ← k × m × (i − 1) + k;
 3 |   if F[i] > 0 then
 4 |   |   if z[i] = ℓ then
 5 |   |   |   z[i] ← z[i − 1] + 1;
 6 |   |   else
 7 |   |   |   z[i] ← z[i] + 1;
 8 |   |   if z ≠ ℓ − 1 then
 9 |   |   |   i ← n;
10 |   |   |   return;
11 |   |   if F[i − 1] > 0 then
12 |   |   |   F[i] ← −F[i − 1];
13 |   |   |   F[i − 1] ← i − 1;
14 |   |   else
15 |   |   |   F[i] ← F[i − 1];
16 |   |   |   F[i − 1] ← 1 − i;
17 |   else
18 |   |   if z[i] = z[i − 1] + 1 then
19 |   |   |   z[i] ← ℓ;
20 |   |   |   if F[i − 1] > 0 then
21 |   |   |   |   F[i] ← F[i − 1];
22 |   |   |   |   F[i − 1] ← i − 1;
23 |   |   |   else
24 |   |   |   |   F[i] ← −F[i − 1];
25 |   |   |   |   F[i − 1] ← 1 − i;
26 |   if F[n] > 0 then
27 |   |   i ← F[n];
28 |   |   F[n] ← n;
29 |   else
30 |   |   i ← −F[n];
31 |   |   F[n] ← −n;
```

Algorithm 1: Loopless

```
 1 begin
 2 |   z[0] ← 0;
 3 |   F[0] ← 1;
 4 |   for j ← 1 to n do
 5 |   |   z[j] ← k × m × (j − 1) + k;
 6 |   |   F[j] ← j;
 7 |   i ← n;
 8 |   Print-Sequence(z[1..n]);
 9 |   while ! (i = 1 and z[1] = k − 1) do
10 |   |   nextZ();
11 |   |   Print-Sequence(z[1..n]);
```

References

1. Amani, M., Nowzari-Dalini, A.: Efficient generation, ranking, and unranking of (k, m)-ary trees in B-order. Bull. Iranian Math. Soc. **45**, 1145–1158 (2019)
2. Chang, Y.-H., Wu, R.-Y., Chang, R.-S., Chang, J.-M.: Improved algorithms for ranking and unranking (k, m)-ary trees in B-order. J. Combin. Optim. (2020). https://doi.org/10.1007/s10878-019-00469-z
3. Du, R.R.X., Liu, F.: (k, m)-Catalan numbers and hook length polynomials for plane trees. Euro. J. Combin. **28**, 1312–1321 (2007)
4. Ehrlich, G.: Loopless algorithms for generating permutations, combinations, and other combinatorial objects. J. ACM **20**, 500–513 (1973)

5. Korsh, J.F., LaFollette, P.: Loopless generation of Gray code for k-ary trees. Inf. Process. Lett. **70**, 7–11 (1999)
6. Pai, K.-J., Chang, J.-M., Wu, R.-Y., Chang, S.-C.: Amortized efficiency of generation, ranking and unranking left-child sequences in lexicographic order. Discrete Appl. Math. **268**, 223–236 (2019)
7. Pallo, J.: Enumerating, ranking and unranking binary trees. Comput. J. **29**, 171–175 (1986)
8. Pallo, J., Racca, R.: A note on generating binary trees in A-order and B-order. Int. J. Comput. Math. **18**, 27–39 (1985)
9. van Baronaigien, D.R.: A loopless Gray-code algorithm for listing k-ary trees. J. Algorithms **35**, 100–107 (2000)
10. Savage, C.D.: A survey of combinatorial Gray codes. SIAM Rev. **39**, 605–629 (1997)
11. Stanley, R.P.: Enumerative Combinatorics, vol. 2. Cambridge University Press, Cambridge (1999)
12. Vajnovszki, V.: On the loopless generation of binary tree sequences. Inf. Process. Lett. **68**, 113–117 (1998)
13. Vajnovszki, V.: Generating a Gray code for P-sequences. J. Math. Model. Algo. **1**, 31–41 (2002)
14. Williamson, S.G.: Combinatorics for Computer Science. Computer Science Press, Rockville (1985)
15. Wu, R.-Y., Chang, J.-M., Chan, H.-C., Pai, K.-J.: A loopless algorithm for generating multiple binary tree sequences simultaneously. Theor. Comput. Sci. **556**, 25–33 (2014)
16. Wu, R.-Y., Chang, J.-M., Peng, S.-L., Liu, C.-L.: Gray-code ranking and unranking on left-weight sequences of binary trees. IEICE Trans. Fund. **E99–A**, 1067–1074 (2016)
17. Wu, R.-Y., Chang, J.-M., Wang, Y.-L.: Loopless generation of non-regular trees with a prescribed branching sequence. Comput. J. **53**, 661–666 (2010)
18. Wu, R.-Y., Chang, J.-M., Wang, Y.-L.: Ranking and unranking of t-ary trees using RD-sequences. IEICE Trans. Inf. Syst. **E94–D**, 226–232 (2011)
19. Wu, R.-Y., Hsu, C.-H., Chang, J.-M.: Loopless algorithms for listing Zaks' sequences in Gray-code order. J. Internet Tech. **15**, 679–684 (2014)
20. Xiang, L., Ushijima, K., Tang, C.: Efficient loopless generation of Gray codes for k-ary trees. Inf. Process. Lett. **76**, 169–174 (2000)
21. Xiang, L., Ushijima, K., Tang, C.: On generating k-ary trees in computer representation. Inf. Process. Lett. **77**, 231–238 (2001)
22. Zaks, S.: Lexicographic generation of ordered trees. Theor. Comput. Sci. **10**, 63–82 (1980)

An LP-Rounding Based Algorithm for a Uniform Capacitated Facility Location Problem with Penalties

Wei Lv[1] and Chenchen Wu[2(✉)]

[1] Computer Science and Technology Department, Tianjin University Renai College,
Tianjin 306136, People's Republic of China
weilupaopao@163.com
[2] College of Science, Tianjin University of Technology,
Tianjin 300384, People's Republic of China
wu_chenchen_tjut@bjut.edu.cn

Abstract. The capacitated constraint is a common variant in the field of combinatorial optimization. In the capcitated facility location problem, there is a limitation of service for each facility, that is, *capacity*. From the view of the approximation algorithm, this variant makes the algorithm difficult to design. Actually, the integral gap of the standard linear program for the capacitated facility location problem is infinite. That is, there is no LP-based constant approximation algorithm. In this work, we consider a special case of the uniform open cost for the capacitated facility location problem. Moreover, all clients are needed to be served. There is another variant of facility location problem called as "robust". In this paper, we also consider one of the robust forms, named cpacitated facility location problem with penalties. We obtain an LP-based 5.732-approximation algorithm for a uniform capacitated facility location problem with penalties.

Keywords: Facility location problem · LP-rounding technique · Approximation algorithm

1 Introduction

1.1 Background

The facility location problem is one of the most important problems in combinatorial optimization. The problem has wide applications in the field of supply chain, computer science, etc. Moreover, the facility location problem is NP-hard problem in generally even the simplest version, uncapcitated facility location problem (UFLP). In UFLP, we are given a facility set and a customer set. The aim is to open some facilities and connect each client to an open facility such that the total cost including facility open cost and connection cost is minimized.

Supported by National Natural Science Foundation of China (No. 11971349).

Moreover, the approximation algorithm is one of the most powerful methods to solve NP-hard problems. An algorithm is α-approximation algorithm if the output by the algorithm is no more than α times optimal value. The ratio of α is called approximation ratio.

In the real world, the capacity of service for a facility is limited which dues to a variant, capacitated facility location problem (CFLP). In the problem, there is a capacity for each facility which is a limitation for serving clients. The demand of a client can be split and served by several facilities. The aim is to open some facilities serving all clients such that the open cost and connection cost is minimized.

Moreover, since the demands of customers are varied, the cost of serving these client is expensive. Thus, this leads to robust facility location problem. One common version is facility location problem with penalty in which there is a penalty cost for each client. The aim is to choose some facilities to open and some clients to be penalized such that the total cost including the open cost, the connection cost and penalty cost is minimized. The other is facility location problem with outliers in which there are at most l clients not to be served. The aim is to choose some facilities to open and no more than l clients not to be served such that the total cost including open cost and connection cost is minimized.

In this work, we consider uniform capacitated facility location problem with penalties. Based on the setting of CFLP, there is a penalty cost for each client. Besides the open cost and connection cost, the penalty cost is a new impact factor of the objective function.

1.2   Literature Reviews

The first constant approximation algorithm for UFLP is given by Shmoys et al. [12]. By LP-rounding technique they obtain a 3.16-approximation algorithm. Combining with dual fitting technique [5], Li [6] introduce nonuniform rounding technique which obtains the best ratio 1.488 for the problem until now. The lower bound of UFLP is 1.463 [11].

Since the integral gap of the standard linear program relaxation of CFLP is infinite, the local-search technique is a common method to design approximation algorithm. By introducing three operations, Aggarwal et al. [1] obtain a 3-approximation algorithm for the special case of uniform capacity. When the capacities are non-uniform, the first constant approximation algorithm is given by Pal et al. [10]. Also using some operations, they obtain a 9-approximation algorithm. Following the paper, the approximation ratio is improved (c.f. [9,14]). The best ratio until now is given by Bansal et al. [3]. An et al. [2] use multicommodity flow obtaining a new linear program relaxation. The approximation ratio is 288. For special cost of the uniform facility cost, Levi et al. [8] give a LP-based 5-approximation algorithm.

For the robust facility location problem, the common versions are facility location problem with penalties and facility location with outliers. We focus on the first variant. Using primal-dual technique, Charikar et al. [4] obtain the first approximation algorithm for the problem, i.e., 3-approximation algorithm. Based on LP-rounding technique, Xu and Xu [13] propose $\left(2 + \frac{2}{e}\right)$-approximation algorithm. The best ratio is given by Li et al. [7] in which they obtain a 1.5148-approximation algorithm by LP-rounding technique based on non-uniform distributed parameter.

The rest of the paper, we will give the elementary knowledge in Sect. 2. Then, we will give the detail algorithm and analysis in Sect. 3. Last, we will give some conclusion in Sect. 4.

2   Preliminaries

2.1   Problem Statement

In the *uniform capacities facility location problem with penalties* (UCFL PWP), we are given a set F of facilities. Each facility $i \in F$ has an open cost f_i and a capacity u_i. We are also given a set D of customers. Each customer $j \in D$ has a demand d_j and a penalty cost p_j. There is a connection cost c_{ij} between facility $i \in F$ and $j \in D$. The aim of the problem is to serve all non-penalized customers by using opened facilities and satisfying the capacity constraints such that the total cost including open cost, connection cost and penalty cost is minimized. We consider the splitable case, that is, the demand of the customer can be served by several opened facilities. Moreover, the *uniform* case means the open cost of all facilities is same, that is, $f_i \equiv f$ for all $i \in F$. Obviously, the problem is a special case of CFLP.

To formulate the problem, 0–1 variable y_i indicates whether facility $i \in F$ is open or not, 0–1 variable z_j indicate whether customer $j \in D$ is penalized or not, and $x_{ij} \in [0,1]$ indicate the fraction that the demand $j \in D$ served by facility $i \in F$. Then, we obtain a linear program for UCFLPWP.

$$IP^* := \min \sum_{i \in F} f_i y_i + \sum_{i \in F} \sum_{j \in D} d_j c_{ij} x_{ij} + \sum_{j \in D} p_j z_j$$

$$\text{s. t.} \sum_{i \in F} x_{ij} + z_j \geq 1, \qquad \forall j \in D, \tag{1}$$

$$x_{ij} \leq y_i, \qquad \forall i \in F, j \in D,$$

$$\sum_{j \in D} d_j x_{ij} \leq u_i y_i, \qquad \forall i \in F,$$

$$0 \leq x_{ij} \leq 1, \qquad \forall i \in F, j \in D,$$

$$y_i, z_j \in \{0,1\}, \qquad \forall i \in F, j \in D.$$

Among the program, the first constraints mean that each customer's demand must to be met, or the cunstomer is penlized; the second constraints mean if facility i offers customer j's demand, facility i must be opened; the third constraints mean that the demands served by a facility i should not beyond the capacity. The linear program relaxation of (1) is

$$LP^* := \min \sum_{i \in F} f_i y_i + \sum_{i \in F} \sum_{j \in D} d_j c_{ij} x_{ij} + \sum_{j \in D} p_j z_j$$

$$\text{s. t. } \sum_{i \in F} x_{ij} + z_j \geq 1, \qquad \forall j \in D, \tag{2}$$

$$x_{ij} \leq y_i, \qquad \forall i \in F, j \in D,$$

$$\sum_{j \in D} d_j x_{ij} \leq u_i y_i, \qquad \forall i \in F,$$

$$y_i \leq 1, \qquad \forall i \in F,$$

$$x_{ij}, y_i, z_j \geq 0, \qquad \forall i \in F, j \in D.$$

The dual program of (2) is

$$\max \sum_{j \in D} \alpha_j - \sum_{i \in F} \zeta_i$$

$$\text{s. t. } \alpha_j \leq d_j c_{ij} + \beta_{ij} + d_j \gamma_i, \qquad \forall i \in F, j \in D, \tag{3}$$

$$\sum_{j \in D} \beta_{ij} \leq f_i + \zeta_i - u_i \gamma_i, \qquad \forall i \in F,$$

$$\alpha_j \leq p_j, \qquad \forall j \in D,$$

$$\alpha_j, \beta_{ij}, \gamma_i, \zeta_i \geq 0, \qquad \forall i \in F, j \in D.$$

2.2   Single Node Capacitated Facility Location Problem

In this work, we need to use a sub-problem called as *single node capacitated facility location problem* (SNCFLP). In SNCFLP, we are given a set of facilities and the corresponding open cost f_i. Each facility has a capacity cap_i. Since there is only one customer, the distance from customer to facility is only about the facility. We mark the distance from customer to facility i is c_i. And the demand of the customer is $\bar{d}$. We need to open some facilities to serve the only customer such that the total cost is minimized. SNCFLP can be represented as the linear mixed integer programming below.

$$\min \sum_i f_i v_i + \sum_i c_i w_i$$

$$\text{s. t. } \sum_i w_i \geq \bar{d}, \tag{4}$$

$$w_i \leq cap_i v_i, \qquad \forall i,$$

$$v_i \in \{0, 1\}, \qquad \forall i,$$

$$w_i \geq 0, \qquad \forall i,$$

In the programming above, 0–1 variable v_i indicates weather facility i is open or not. w_i indicates the fraction of the customer's demand served by the facility

i. The first constraint means the summation of demands that all the facilities offer to customer should be more than $\bar{d}$. The second constraint means that the demand facility i offers to customer should not exceed the capacity of the facility. The linear programming relaxation of SNCFLP is

$$\min \; \sum_i f_i v_i + \sum_i c_i w_i$$

$$\text{s. t.} \; \sum_i w_i \geq \bar{d}, \tag{5}$$

$$w_i \leq cap_i v_i, \qquad \forall i,$$

$$v_i \leq 1, \qquad \forall i,$$

$$v_i, w_i \geq 0, \qquad \forall i.$$

Without loss of generality, we assume $v_i := \frac{w_i}{cap_i}$ in the optimal solution (v, w) of (5). Then, the program (5) is equivalent to

$$\min \; \sum_i \left(\frac{f_i}{cap_i} + c_i \right) w_i$$

$$\text{s. t.} \; \sum_i w_i \geq \bar{d}, \tag{6}$$

$$w_i \leq cap_i, \qquad \forall i,$$

$$w_i \geq 0, \qquad \forall i.$$

The optimal solution of (6) is easily to verify by a simple greedy algorithm. That is, order all facilities with non-decreasing order of $f_i/cap_i + c_i$. Following the above ordering, set w_i as cap_i until the summation exceed $\bar{d}$. Then, let $w_i := \bar{d} - \sum_{i' < i} w_i$ for the last facility i and $v_i := w_i/cap_i$. For the rest facilities, let $v_i = w_i = 0$. It is easy to verify (v, w) is an optimal solution. Moreover, there is at most one facility i opened for $0 < v_i < 1$.

3   Algorithm and Analysis

In our algorithm, we use the standard LP-rounding framework. First, we solve the linear program relaxation to obtain a fractional optimal solution (x^*, y^*, z^*) and the corresponding dual optimal solution $(\alpha^*, \beta^*, \gamma^*, \zeta^*)$. Second, we define the penalized customer set by a threshold. For the left customers, we divide these facilities into different customer-centric clusters. Third, in each cluster, we open some facilities and connect non-penalized customers to open facilities.

3.1   Clustering

Given a parameter $\phi \in (0, 1)$, we decide the customer with $z_j^* \geq \phi$ to be penalized, i.e., let the penalized customer set $\hat{P} = \{j \in D : z_j^* \geq \phi\}$. The rest customers need to be served. Denote D' as the set of the served customers, that is, $D' = D \setminus \hat{P}$. Then, we partition F into several clusters. In each cluster, there is a center $j \in D'$ and several facilities in F. The procedure is as follows. Denote C' as the set of centers, N_j as the facilities in the cluster centered by j, U as the set of candidate centers.

- Step 1.1. Initially, set $C' = \emptyset$, $N_j := \{i : x_{ij}^* > 0\}$, and $U := D'$.
- Step 1.2. Choose a customer in U with minimal $\frac{\alpha_j^*}{d_j}$, i.e.,

$$j' = \arg\min_{j \in U} \frac{\alpha_j^*}{d_j}.$$

  Then, set $C' = C' \cap \{j'\}$, and $U = U \setminus \{j'\}$.
- Step 1.3. For all $j \in U$, define $N_j^{far} := \{i \in N_j : c_{ij} > c_{ij'}\}$. The facilities in N_j^{far} is relatively far away from j. Thus, update $N_j = N_j \setminus \left(N_{j'} \cup N_j^{far} \right)$.
- Step 1.4. Update the set of candidate centers,

$$U = U \setminus \left\{ j \in U : \sum_{i \in N_j} x_{ij}^* < \frac{1}{2}(1 - \phi) \right\}.$$

  Go back to Step 1.2 until U is empty.
- Step 1.5. At the last step, there are some facilities not clustered. For all $i \notin \cup_{j' \in C'} N_{j'}$, denote $j^*(i)$ as the closest center in C' for i. Then, add i to $N_{j^*(i)}$. For convenience, we extend $j^*(i)$ for all facilities in F. Denote $j^*(i)$ as the center of the cluster which facility $i \in F$ belongs to.

In the following of the paper, N_j for $j \in C'$ denotes the set when j is added into C' and N_j for $j \notin C'$ denotes the set when j is deleted from U.

Lemma 1. *The penalty cost for the penalized set $\hat{P}$ is no more than $\frac{1}{\phi} \sum_{i \in \hat{P}} p_j z_j^*$.*

3.2 Opening Facilities

In this section, we choose some facilities to open in each cluster $(\{j'\}, N_{j'})$.

Step 2.1. For the faiclity $i \in N_{j'}$ with $y_i^* = 1$, open these facilities. That is, $\hat{y}_i = 1$ for all $\{i : y_i^* = 1\}$.

Step 2.2. Define SNCFLP as follows. The facility set is $F_{j'} := \{i : 0 < y_i^* < 1\}$. The only client is the center j'. The demand of j' is the total amount served by the facilities, that is, $\bar{d}_{j'} = \sum_{i \in F_{j'}} \sum_{j \in D \setminus \hat{P}} d_j x_{ij}^*$. The capacity cap_i for each facilities i in $F_{j'}$ is still u_i. The connection cost c_i is $c_{ij'}$.

Step 2.3. Using the greedy algorithm, we obtain a optimal solution $(v^{j'}, w^{j'})$. We open all facilities with $v_i^{j'} > 0$. That is, $\hat{y}_i = 1$ with all facilities in $\{i \in F_{j'} : v_i^{j'} > 0\}$.

Note in Step 2.3, there is only one fractional open facility is open. Denote the special facility as i_s. We estimate the cost of the facility.

Lemma 2

$$f_{i_s} \leq \frac{2}{1-\phi} \sum_{i \in N_{j'}} f_i y_i^*.$$

Proof. By the procedure of clustering, $\sum_{i \in N_{j'}} x_{ij'}^* \geq \frac{1}{2}(1 - \phi)$. Because of the uniform open cost, we have

$$f_{i_s} \leq f_{i_s} \frac{2}{1-\phi} \sum_{i \in N_{j'}} x_{ij'}^* \leq \frac{2}{1-\phi} \sum_{i \in N_{j'}} f_i y_i^*.$$

Since we use SNCFLP as a sub-problem, the fractional optimal value $Cost_{sncflp}^{j'}$ of SNCFLP in cluster $(j', N_{j'})$ will be used to estimate the cost of $(\hat{x}, \hat{y})$.

Lemma 3. *In the cluster* $(\{j'\}, N_{j'})$, *we have*

$$Cost_{sncflp}^{j'} \leq \sum_{i \in F_{j'}} f_i y_i^* + \sum_{j \in D \setminus \hat{P}} \sum_{i \in F_{j'}} d_j c_{ij'} x_{ij}^*.$$

Proof. If we let $v_i := y_i^*$ and $w_i = \sum_{j \in D \setminus \hat{P}} d_j x_{ij}^*$, the solution (v, w) is a feasible solution for SNCFP. Indeed,

$$\sum_{i \in F_{j'}} w_i = \sum_{i \in F_{j'}} \sum_{j \in D \setminus \hat{P}} d_j x_{ij}^* = \bar{d}_{j'}.$$

$$w_i = \sum_{j \in D \setminus \hat{P}} d_j x_{ij}^* \leq u_i y_i^* = u_i v_i.$$

Thus,

$$Cost_{sncflp}^{j'} \leq \sum_{i \in F_{j'}} f_i v_i + \sum_{i \in F_{j'}} c_{ij'} w_i$$

$$= \sum_{i \in F_{j'}} f_i y_i^* + \sum_{j \in D \setminus \hat{P}} \sum_{i \in F_{j'}} d_j c_{ij'} x_{ij}^*.$$

3.3   Connecting Customers

When we get the open facilities, the connection can be obtained by solving a transportation problem. In order to analyze easily, we construct a feasible way of connection in each cluster $\{j', N_{j'}\}$.

Step 3.1. For the facility $i \in N_{j'} \setminus F_{j'}$, keep the fractional connections of fractional optimal solution. That is, $\hat{x}_{ij} = x_{ij}^*$.

Step 3.2. For the facility $i \in F_{j'}$, solve the following system to obtain feasible solution $\hat{x}_{ij}$,

$$\left\{ \hat{x}_{ij} : \sum_{i \in F_{j'}} \hat{x}_{ij} = \sum_{i \in F_{j'}} x_{ij}^*, \forall j \in D \setminus \hat{P}; \sum_{j \in D \setminus \hat{P}} d_j \hat{x}_{ij} = w_i^{j'}, \forall i \in F_{j'} \right\}.$$

Since the complementary relaxation condition of linear program, the following lemmas hold in [8].

Lemma 4. *For a client* $j \in D$ *and the fractionally connected facility* i *in* (x^*, y^*, z^*), *that is,* $x_{ij}^* > 0$,

Case 1: $j \in C'$ *If* $i \in N_j$, $c_{ij^*(i)} \leq c_{ij}$. *If* $i \in \bar{N}_j$, $c_{ij^*(i)} \leq \frac{\alpha_j^*}{d_j}$.

Case 2: $j \notin C'$ *If* $i \in N_j$, $c_{ij^*(i)} \leq c_{ij} + c_{i^*(j)j} + \frac{\alpha_j^*}{d_j}$ *where* $i^*(j) = \arg\min_{i \in \bar{N}_j} \{c_{ij}\}$.

If $i \in \bar{N}_j$, $c_{ij^*(i)} \leq \frac{\alpha_j^*}{d_j}$.

Lemma 5. *For the facility* i *with* $y_i^* = 1$ *and corresponding connection cost is no more than*

$$\sum_{j \in D} \sum_{i \in F : y_i^* = 1} \alpha_j^* x_{ij}^* - \sum_{i \in F} \zeta_i^*.$$

Now we are ready to estimate the left open cost and connection cost.

Lemma 6. *In the cluster* $(j', N_{j'})$, *the total cost including open cost and corresponding connection cost for the fractional open facilities in* (x^*, y^*, z^*) *is no more than*

$$\left(1 + \frac{2}{1 - \phi}\right) \sum_{i \in N_{j'}} f_i y_i^* + \sum_{j \in D \setminus \hat{P}} \sum_{i \in F_{j'}} d_j c_{ij} x_{ij}^* + 2 \sum_{j \in D \setminus \hat{P}} \sum_{i \in F_{j'}} d_j c_{ij'} x_{ij}.$$

Proof. Since the triangle inequalities and the define of $\hat{x}_{ij}$, we have

$$\sum_{j \in D \setminus \hat{P}} \sum_{i \in F_{j'}} d_j c_{ij} \hat{x}_{ij} \leq \sum_{i \in F_{j'}} \sum_{j \in D \setminus \hat{P}} d_j c_{ij'} \hat{x}_{ij} + \sum_{i \in F_{j'}} \sum_{j \in D \setminus \hat{P}} d_j c_{jj'} \hat{x}_{ij}$$

$$\leq \sum_{i \in F_{j'}} c_{ij'} w_i^j + \sum_{j \in D \setminus \hat{P}} \sum_{i \in F_{j'}} d_j c_{ij'} x_{ij}^* + \sum_{j \in D \setminus \hat{P}} \sum_{i \in F_{j'}} d_j c_{ij'} x_{ij}^*.$$

To sum up, it can be obtained by combining lemma 3 and Lemma 2

$$\sum_{i \in F_{j'}} f_i \hat{y}_i + \sum_{j \in D \backslash \hat{P}} \sum_{i \in F_{j'}} d_j c_{ij} \hat{x}_{ij}$$

$$\leq \sum_{i \in F_{j'}} f_i v_i^{(j')} + \frac{2}{1-\phi} \sum_{i \in N_{j'}} f_i y_i^* + \sum_{i \in F_{j'}} c_{ij'} w_i^{(j')} + \sum_{j \in D \backslash \hat{P}} \sum_{i \in F_{j'}} d_j c_{ij} x_{ij}^*$$

$$+ \sum_{j \in D \backslash \hat{P}} \sum_{i \in F_{j'}} d_j c_{ij'} x_{ij}^*$$

$$= Cost_{sncflp}^{j'} + \frac{2}{1-\phi} \sum_{i \in N_{j'}} f_i y_i^* + \sum_{j \in D \backslash \hat{P}} \sum_{i \in F_{j'}} d_j c_{ij} x_{ij}^* + \sum_{j \in D \backslash \hat{P}} \sum_{i \in F_{j'}} d_j c_{ij'} x_{ij}^*$$

$$\leq \sum_{i \in F_{j'}} f_i y_i^* + \sum_{j \in D \backslash \hat{P}} \sum_{i \in F_{j'}} d_j c_{ij'} x_{ij}^* + \frac{2}{1-\phi} \sum_{i \in N_{j'}} f_i y_i^* + \sum_{j \in D \backslash \hat{P}} \sum_{i \in F_{j'}} d_j c_{ij} x_{ij}^*$$

$$+ \sum_{j \in D \backslash \hat{P}} \sum_{i \in F_{j'}} d_j c_{ij'} x_{ij}^*$$

$$\leq \left(1 + \frac{2}{1-\phi}\right) \sum_{i \in N_{j'}} f_i y_i^* + \sum_{j \in D \backslash \hat{P}} \sum_{i \in F_{j'}} d_j c_{ij} x_{ij}^* + 2 \sum_{j \in D \backslash \hat{P}} \sum_{i \in F_{j'}} d_j c_{ij'} x_{ij}^*.$$

Then, we can estimate the approximation ratio.

Theorem 1. *There is a 5.732-approximation algorithm for UCFLPWP.*

Proof. Using Lemma 6, we sum over all $j' \in C'$ to get

$$\sum_{i \in F : y_i^* < 1} f_i \hat{y}_i + \sum_{j \in D \backslash} \sum_{i \in F : y_i^* < 1} d_j c_{ij} \hat{x}_{ij}$$

$$\leq \left(1 + \frac{2}{1-\phi}\right) \sum_{i \in F} f_i y_i^* + \sum_{j \in D \backslash \hat{P}} \sum_{i \in F : y_i^* < 1} d_j c_{ij} x_{ij}^*$$

$$+ 2 \sum_{j \in D \backslash \hat{P}} \sum_{i \in F : y_i^* < 1} d_j c_{ij^*(i)} x_{ij}^*.$$

Considering the last item on the right side of the above equation, the distance from i to $j^*(i)$ need to be estimate. Through Lemma 4 and Lemma 5.

$$2 \sum_{j \in D \backslash \hat{P}} \sum_{i \in F: y_i^* < 1} d_j c_{ij^*(i)} x_{ij}^*$$

$$\leq 2 \sum_{j \in C'} \sum_{i \in F: y_i^* < 1} d_j c_{ij^*(i)} x_{ij}^* + 2 \sum_{D \backslash (\hat{P} \cup C')} \sum_{i \in F: y_i^* < 1} d_j c_{ij^*(i)} x_{ij}^*$$

$$\leq 2 \left(\sum_{j \in C'} \sum_{i \in \bar{N}_j: y_i^* < 1} \alpha_j^* x_{ij}^* + \sum_{j \in C'} \sum_{i \in N_j} d_j c_{ij} x_{ij}^* + \sum_{j \in D \backslash (\hat{P} \cup C')} \sum_{i \in \bar{N}_j: y_i^* < 1} \alpha_j^* x_{ij}^* \right.$$

$$\left. + \sum_{j \notin D \backslash (\hat{P} \cup C')} \sum_{i \in N_j: y_i^* < 1} (d_j(c_{ij} + c_{i^*(j)j}) x_{ij}^* + \alpha_j^* x_{ij}^*) \right)$$

$$\leq 2 \sum_{j \in D \backslash \hat{P}} \sum_{i \in F: y_i^* < 1} \alpha_j^* x_{ij}^* + 2 \sum_{j \in D \backslash \hat{P}} \sum_{i \in N_j} d_j c_{ij} x_{ij}^*$$

$$+ 2 \sum_{j \in D \backslash (\hat{P} \cup C')} \sum_{i \in N_j} d_j c_{i^*(j)j} x_{ij}^*. \tag{7}$$

When $j \in D \backslash (\hat{P} \cup C')$, we have $\sum_{i \in N_j} x_{ij}^* < \frac{1}{2}(1 - \phi)$. Thus,

$$\frac{2}{1 - \phi} \sum_{i \in N_j} c_{i^*(j)j} x_{ij}^* \leq c_{i^*(j)j} = \min_{i \in \bar{N}_j} \{c_{ij}\}$$

$$\leq \frac{\sum_{i \in \bar{N}_j} c_{ij} x_{ij}^*}{\sum_{i \in \bar{N}_j} x_{ij}^*} < \frac{2}{1 - \phi} \sum_{i \in \bar{N}_j} c_{ij} x_{ij}^*.$$

Thus, the last term of (7) is no more than $2 \sum_{i \in \bar{N}_j} c_{ij} x_{ij}^*$. Then, the cost of the

$$\sum_{i \in F: y_i^* < 1} f_i \hat{y}_i + \sum_{j \in D \backslash \hat{P}} \sum_{i \in F: y_i^* < 1} d_j c_{ij} \hat{x}_{ij}$$

$$\leq \left(1 + \frac{2}{1 - \phi}\right) \sum_{i \in F} f_i y_i^* + \sum_{j \in D \backslash \hat{P}} \sum_{i \in F} d_j c_{ij} x_{ij}^* + 2 \sum_{j \in D \backslash \hat{P}} \sum_{i \in F: y_i^* < 1} \alpha_j^* x_{ij}^*$$

$$+ 2 \sum_{j \in D \backslash \hat{P}} \sum_{i \in N_j} d_j c_{ij} x_{ij}^* + 2 \sum_{j \in D \backslash \hat{P}} \sum_{i \in \bar{N}_j} d_j c_{ij} x_{ij}^*$$

$$\leq \left(1 + \frac{2}{1 - \phi}\right) \sum_{i \in F} f_i y_i^* + 3 \sum_{j \in D \backslash \hat{P}} \sum_{i \in F} d_j c_{ij} x_{ij}^* + 2 \sum_{j \in D \backslash \hat{P}} \sum_{i \in F: y_i^* < 1} \alpha_j^* x_{ij}^*.$$

Overall, combined Lemma 5, we have

$$\sum_{i \in F} f_i \hat{y}_i + \sum_{j \in D \setminus \hat{P}} \sum_{i \in F} d_j c_{ij} \hat{x}_{ij} + \sum_{j \in \hat{P}} p_j$$

$$\leq \left(\sum_{j \in D} \sum_{i \in F : y_i^* = 1} \alpha_j^* x_{ij}^* - \sum_{i \in F} \zeta_i^* \right) + 2 \sum_{j \in D \setminus \hat{P}} \sum_{i \in F : y_i^* < 1} \alpha_j^* x_{ij}^*$$

$$+ \left(1 + \frac{2}{1 - \phi} \right) \sum_{i \in F} f_i y_i^* + 3 \sum_{j \in D \setminus \hat{P}} \sum_{i \in F} c_{ij} x_{ij}^* + \frac{1}{\phi} \sum_{j \in \hat{P}} p_j z_j^*$$

$$\leq 2 \left(\sum_{j \in D} \sum_{i \in F} \alpha_j^* - \sum_{i \in F} \zeta_i^* \right) + \max \left\{ 1 + \frac{2}{1 - \phi}, 3, \frac{1}{\phi} \right\} LP^*.$$

Setting $\phi := 2 - \sqrt{3}$, we get

$$\sum_{i \in F} f_i \hat{y}_i + \sum_{j \in D \setminus \hat{P}} \sum_{i \in F} d_j c_{ij} \hat{x}_{ij} + \sum_{j \in \hat{P}} p_j$$

$$\leq 2 \left(\sum_{j \in D} \alpha_j^* - \sum_{i \in F} \zeta z_i^* \right) + (2 + \sqrt{3}) LP^*$$

$$\leq (4 + \sqrt{3}) LP^* \leq (4 + \sqrt{3}) IP^* \approx 5.732 IP^*.$$

4   Conclusions

In this paper, we consider a form of robust capacitated facility location problem, i.e., capcitated facility location problem with penalty in which the facility cost is uniform. We obtain a 5.732-approximation algorithm based on LP-rounding framework. In analysis of this paper, some inequalites are not tight. We believe the approximation ratio can be improved by more careful analysis. Moreover, capacities version is a local constraint for facility location problem. It is interesting to study global constraint for facility location problem, for example capcitated k-facility location problem.

References

1. Aggarwal, A., Louis, A., Bansal, M., Garg, N., Gupta, N., Gupta, S., Jain, S.: A 3-approximation algorithm for the facility location problem with uniform capacities. Math. Program. **141**, 527–547 (2013)
2. An, H.C., Singh, M., Svensson, O.: LP-based algorithms for capacitated facility location. SIAM J. Comput. **46**, 272–306 (2017)
3. Bansal, M., Garg, N., Gupta, N.: A 5-Approximation for Universal Facility Location. In: Ganguly S., Pandya P.K. (eds.) IARCS 2018, LIPIcs, vol. 122, pp. article No. 24. Schloss Dagstuhl - Leibniz-Zentrum für Informatik, Ahmedabad (2018). https://doi.org/10.4230/LIPIcs.FSTTCS.2018.24

4. Charikar, M., Khuller, S., Mount, D.M., Narasimhan, G.: Algorithms for facility location problems with outliers. In: Proceedings of the 13th Annual ACM-SIAM Symposium on Discrete Algorithms, pp. 642–651. ACM/SIAM, Washington, DC (2001)
5. Jain, K., Mahdian, M., Markakis, E., Saberi, A., Vazirani, V.V.: Greedy facility location algorithms analyzed using dual fitting with factor-revealing LP. J. ACM **50**, 795–824 (2003)
6. Li, S.: A 1.488-approximation algorithm for the uncapacitated facility location problem. Inf. Comput. **222**, 45–58 (2013)
7. Li, Y., Du, D., Xiu, N., Xu, D.: Improved approximation algorithms for the facility location problems with linear/submodular penalty. Algorithmica **73**, 460–482 (2015)
8. Levi, R., Shomys, D.B., Swamy, C.: LP-based approximation algorithms for capacitated facility location. Math. Program. **131**, 365–379 (2012)
9. Mahdian, M., Pál, M.: Universal facility location. In: Di Battista, G., Zwick, U. (eds.) ESA 2003. LNCS, vol. 2832, pp. 409–421. Springer, Heidelberg (2003). https://doi.org/10.1007/978-3-540-39658-1_38
10. Pal, M., Tardos, E., Wexler, T.: Facility location with nonuniform hard capacities. In: Proceedings of 42nd Annual Symposium on Foundations of Computer Science, pp. 329–338, IEEE Computer Society, Las Vegas, Nevada, USA (2001)
11. Guha, S., Khuller, S.: Greedy strike back: improved facility location algorithms. J. Algorithms **31**, 228–248 (1999)
12. Shmoys, D.B.: Tardos E, Aardal K I, Approximation algorithms for facility location problems. In: Proceedings of the 29th Annual ACM Symposium on the Theory of Computing, pp. 265–274. ACM, Texas (1997)
13. Xu, G., Xu, J.: An improved approximation algorithm for uncapacitated facility location problem with penalties. J. Combinat. Optim. **17**, 424–436 (2008)
14. Zhang, J., Chen, B., Ye, Y.: A multi-exchange local search algorithm for the capacitated facility location problem. Math. Oper. Res. **30**, 389–403 (2003)

Author Index

Printed in the United States
By Bookmasters